Finite Element Method with Applications in Engineering

Finite Element Method with Applications in Engineering

YINGXIAO SONG

ARCLER
PRESS

www.arclerpress.com

Finite Element Method with Applications in Engineering

YINGXIAO SONG

Arcler Press
2010 Winston Park Drive,
2nd Floor
Oakville, ON L6H 5R7
Canada
www.arclerpress.com
Tel: 001-289-291-7705
001-905-616-2116
Fax: 001-289-291-7601
Email: orders@arclereducation.com

ISBN: 978-1-77361-234-8 (Hardcover)

ABOUT THE AUTHOR

YINGXIAO SONG

Yingxiao Song has master degree from Inner Mongolia University of Technology. She has taken part in many projects, which are related to the mechanical design, image process, and biomechanics. She is a current PhD student from University of Nebraska-Lincoln. Her current research topic is micro forming.

TABLE OF CONTENTS

PREFACE

Finite element method is a powerful analysis tool in engineering field. While how it reflected in the industry field is still very unfamiliar for many people. This book presents a review of finite element method application in industry field. First three chapter introduced how finite element method works. Chapter 1 is a brief introduction of FEM's history and basic theory. Chapter 2 illustrates the mathematic models of FEM. Chapter 3 introduces three most popular commercial software. Chapter 4 to 7 shows the FEM's application in solid mechanics, structures, heat transfer and manufactory. Specific examples are provided in each topic. Thanks to the people who works so hard on this field, enjoy!

CHAPTER 1

INTRODUCTION OF FINITE ELEMENT METHOD

DEVELOPMENT OF FINITE ELEMENT METHOD

The origin of Finite Element Method can trace back to 1950, when M.J. (Jon) Turner generalized and perfected the Direct Stiffness Method, and forcefully got Boeing to commit resources to it while other aerospace companies were mired in the Forced Method. In 1952, he oversaw the development of the first continuum based finite elements. Then the development of Finite Element Method went through four stages: the pioneers, golden age, consolidation and back to basic.

J. H. Argyris, R. W. Clough, H. C. Martin, and O. C. Zienkiewicz popularize the Finite Element Method after Turner. They transferred this technology from aerospace industry to a wider range of engineering applications. In 1956, Turner, Clough, Martin and Topp published Stiffness and Deflection Analysis of Complex Structure in Journal of the Aeronautical Sciences. It is recognized as the start of the current FEM. Along with Argyris's serial, Przemieniecki's textbook, Gallagher's survey, the panoramic picture of this period is available. All of the people mentioned here are from aerospace industry, since FEM has the requirements for digital computation, only large industrial companies were able to afford mainframe computers during the 1950s.

Figure 1.1 Dr. John H. Argyris (8.19.1913-4.2.2004), one of the creators of the finite element method.

1962 to 1972 is the golden age of Finite Element Method. Melosh published an influential paper, marking the confluence of three lines of research: Argyris' dual formulation of energy methods, the Direct Stiffness Method and Convergence. A two-step interpretation of FEM emerges: discrete elements approximate continuum models, which in turn approximate real structures. FEM begins to expand into Civil Engineering through Clough's Boeing-Berkeley connection and had been baptized. Reading Fraejis De Veubeke's famous article one can sense the ongoing change in perspective opened up by the variational framework. The first book devoted to FEM appears in 1967. Applications to nonstructural problems had started in 1965, and were treated in some depth by Martin and Carey.

Textbooks by Hughes and Bathe reflect the technology of the consolidation period. Hybrid and mixed formulations record steady progress. Assumed strain formulations appear. A booming activity in error estimation and mesh adaptively is fostered by better understanding of the mathematical foundations. By early 1980s, more approaches come on the scene, notably the free formulation, orthogonal hourglass control, assumed natural strain methods, stress hybrid models in natural coordinates. As well as variants and derivatives of those approaches. Two more recent trends can be noted: increased abstracion on the mathematical side, and canned recipes for running commercial software on the physical side.

As used today, FEM represents the confluence of three ingredients: Matrix Structure Analysis, Variational approximation theory, and the digital computer, which came together in the early 1950. The main history milestones are given in the follows:

1941-1942 – Hrennikoff and Courant developed mesh discretization methods for solving elasticity and structural analysis problems in civil and aeronautical engineering.

1956 – Ray W. Clough published the first paper on the finite element method (FEM). "Finite elements" term coined in 1960 article.

1959 – General Motors and IBM build computer system DAC-1 (Design Augmented by Computers) to facilitate cars design.

1960 – William Fetter from Boeing coins the term "computer graphic" for his human factors cockpit drawings.

1965 – NASTRAN (NASA Structural Analysis) is developed as structural analysis solver tool.

1977 – FIESTA, the first professional FEM p-version code, initiated by Alberto Peano from ISME.

1982 – PROBE, developed by Barna Szabo and Kent Myers, 1st 'industrial' implementation of p-version FEA for research and aerospace applications.

1987 – MECHANICA, developed by RASNA Corp.

2001 – P-version FEM proven to be the most efficient for Plasticity by A. Duster.

2006 – ASME guide for Verification and Validation in Computational Solid Mechanics is released.

2008 – NASA released a Standard for development of Models and Simulations.

2012 – Barna Szabo and Ricardo Actis introduce "Simulation governance: New technical requirements for software tools in computational solid mechanics".

2013 – SimScale officially released the world's first cloud-based 3D simulation platform, offering a comprehensive set of linear and non-linear FEA capabilities for structural mechanics, fluid dynamics, and thermodynamics.

THEORY OF FINITE ELEMENT

The finite element method is a numerical procedure for solving engineering problems. The concept of FEM will be partly illustrated through a truly ancient problem: find the perimeter L of circle of diameter d. Since, this is equivalent to obtaining a numerical value of. Draw a circle of radius r and diameter as in Figure 1.2.1. Inscribe a regular polygon of n sides, where n=8 in Figure 1.2.1(b). Rename polygon sides as elements and vertices as nodes. Label nodes with integer 1…8. Extract a typical element, say that joining nodes 4-5, as shown in Figure. This is an instance of the generic element i-j pictured in Figure 1.2.1(d). The element length is. Since all elements have the same length, the polygon perimeter is, whence the approximation to is. As can be seen the convergence to is fairly slow. However, the sequence can be transformed by Wynn's algorithm into that shown in the third column.

Some key ideas behind the FEM can be identified in this example. The circle, viewed as a source mathematical object, is replaced by polygons. These are discrete approximations to the circle. The sides, renamed as elements, are specified by their end nodes. Elements can be separated by disconnecting nodes, a process called disassembly in the FEM. Upon disassembly a generic element can be defined, independently of the original circle, by the segment that connects two nodes i and j. The relevant element property: side length, can be computed in the generic element independently of the others, a property called local support on the FEM. The target property: polygon perimeter, is obtained by reconnecting n elements and adding up their length; the corresponding steps in the FEM being assembly and solution, respectively. There is of course nothing magic about the circle; the same technique can be used to rectify any smooth plane curve.

This example has been offered in the FEM literature to adduce that finite element ideas can be traced to Egyptian mathematicians from circa 1800 B.C., as well as Archimedes' famous studies on circle rectification by 250 B.C. But comparison with the modern FEM shows this to be a stretch. The example does not illustrate the concept of degrees of freedom, conjugate quantities and local-global coordinates. It is guilty of circular reasoning: the compact formula uses the unknown in the right hand side. Reasonable people would argue that a circle is a simler object than, say, a 128-sided polygon. Despite these flaws the example is useful in one respect: showing a fielder's choice in the replacement of one mathematical object by another. This is at the root of the simulation process.

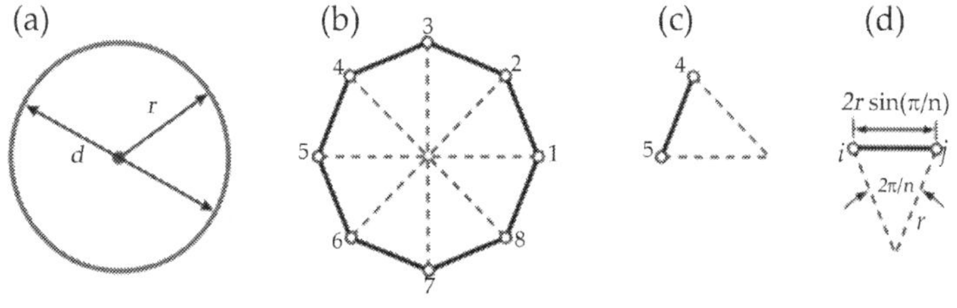

n	$\pi_n = n\sin(\pi/n)$	Extrapolated by Wynn-ϵ	Exact π to 16 places
1	0.000000000000000		
2	2.000000000000000		
4	2.828427124746190	3.414213562373096	
8	3.061467458920718		
16	3.121445152258052	3.141418327933211	
32	3.136548490545939		
64	3.140331156954753	3.141592658918053	
128	3.141277250932773		
256	3.141513801144301	3.141592653589786	3.141592653589793

Figure 1.2 the "find" problem treat with FEM concepts: (a) continuum object, (b) a discrete approximation by inscribed regular polygons, (c) disconnected element, (d) generic element. The table at the bottom shows rectification of circle by inscribed polygons ("Archimedes FEM").

Use linear elastic behavior as example, there are six steps in solving these problems by finite element analysis. They can be summarized as follows:

1. Discretization the domain

This step involves subdividing the domain into elements and nodes. For discrete systems, this step is unnecessary. However, for continuous systems like plates and shells this step becomes very important and the answers obtained are only approximate. In this case, the accuracy of the solution depends on the discretization used.

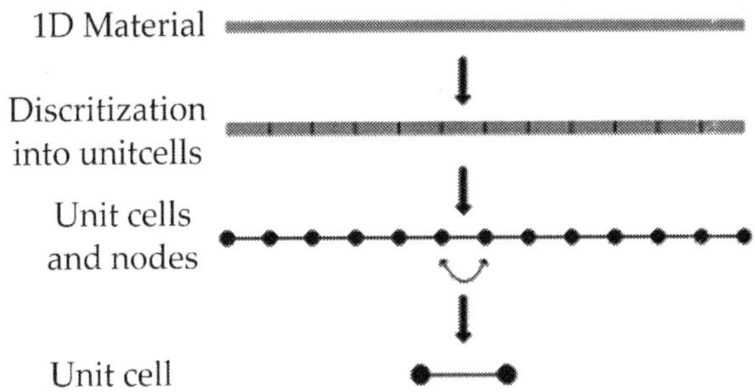

Figure 1.3. Discretization of 1D material, the continuous bar is discretized into finite unicell, each unicell has two nodes, the image is from Active Aeroelasticity and Structures Research Laboratory, University of Michigan.

The discretization of the domain is the first and perhaps the most important step in any finite element analysis because the manner in which the domain is discretized will affect the computer storage requirements, the computation time, and the accuracy of the numerical results. The subdomains are usually referred to as the elements. For a 1D domain which is actually a straight of curved line, the elements are often short line segments interconnected to form the original line [Figure 1.2.3(a)]. For a 2D domain, the elements are usually small triangles and rectangles [Figure 1.2.3(b)]. The rectangular elements are, of course, best suited for discretizing rectangular regions, while the triangular ones can be used for irregular regions. In a 3D solution, the domain may be subdivided into tetrahedra, triangular prisms, or rectangular bricks[Fig1.2.3(c)], among which the tetrahedra are the simplest and best suited for arbitrary-volume domains.

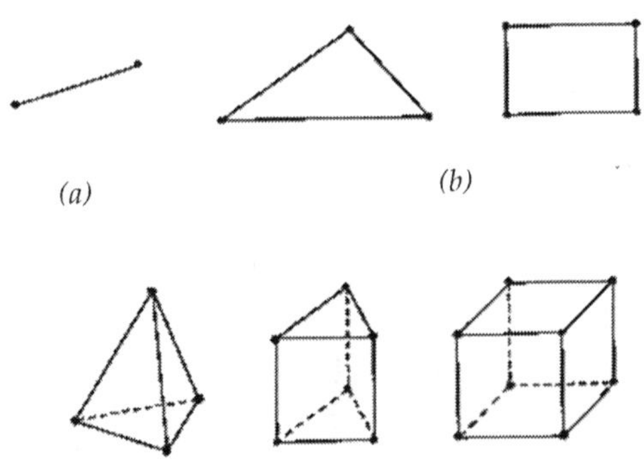

Figure1.4 Basic finite elements. (a) 1D (b) 2D (c) 3D

Note that the linear line segments, triangles, and tetrahedra are the basic one-, two-, and three-dimensional elements. Figure1.2.4 shows the finite element discretization of a 2- and 3- dimensional domain.

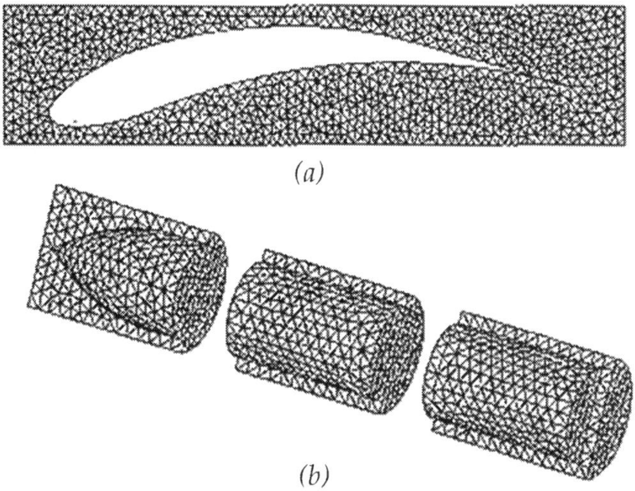

(a)

(b)

Figure1.5 Examples of finite element discretization (a) 2-D with triangular elements (b) 3-D with tetrahedra elements

The discretization of the domain is usually considered as a preprocessing task because it can be completely separated from the other steps. Many well-developed finite element program packages have the capability of subdividing an arbitrarily shaped line, surface, and volume into the corresponding elements and also provide the optimized global numbering.

Writing the element stiffness matrices, the element stiffness equations need to be written for each element in the domain and assembling the global stiffness matrix, this will be done using the direct stiffness approach.

Applying the boundary conditions, like support and applied loads and displacements.

Solving the equations, this will be done by partitioning the global stiffness matrix and then solving the resulting equations using Gaussian elimination.

Post-processing, to obtain additional information like the reactions and element forces and stresses.

Industry Application

Currently, Finite Element Methods are so embedded within the world of engineering that even an undergraduate mechanical engineering student would have already worked with a FEA package, such as ANSYS or ABAQUS, during his or her academic studies.

The first step in the solution of a problem is the identification of the problem itself. Hence, before we can analyze a structure we must ask ourselves the following questions: Which are the more relevant physical phenomena? Is the problem of static or dynamic nature? Are the kinematics or the material properties linear or non-linear? Which are the key results requested? What is the level of accuracy sought? The answers to these questions are essential for selecting a structural model and the adequate computational method.

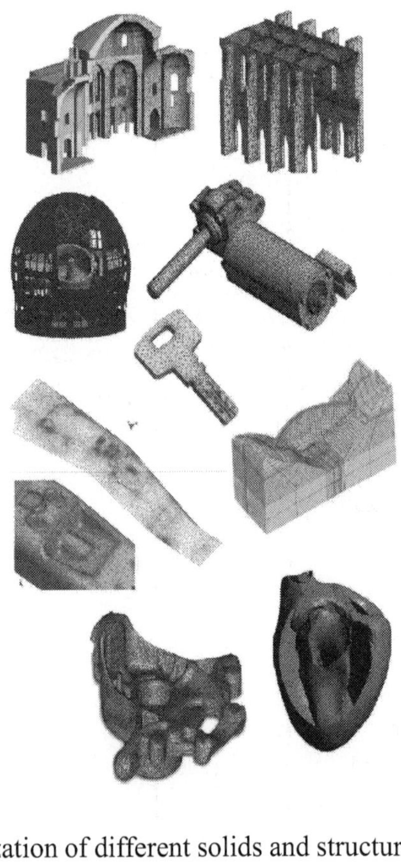

Figure. 1.6 Discretization of different solids and structures with finite elements

The next step in the structural analysis sequence is the definition of a numerical method, such as the FEM. The application of the FEM invariably requires its implementation in a computer code. The analysis of a structure with the FEM implies feeding the code with quantitative in-formation on the mechanical properties of the materials, the boundary conditions and the applied loads (the physical parameters) as well as the features of the discretization chosen (i.e. element type, mesh size, etc). The outcome of this process is what we call a computational model.

Figure 1.7 shows the actual image of a car panel, the geometrical definition of the panel surface by means of NURBS (non-uniform rational B-splines) patches using computer-aided design (CAD) tools. The discretization of the surface by a mesh of 3-noded shell triangles and some numerical results of the FEM analysis. A similar example of the FEM analysis of an office building is shown in Figure 1.8.

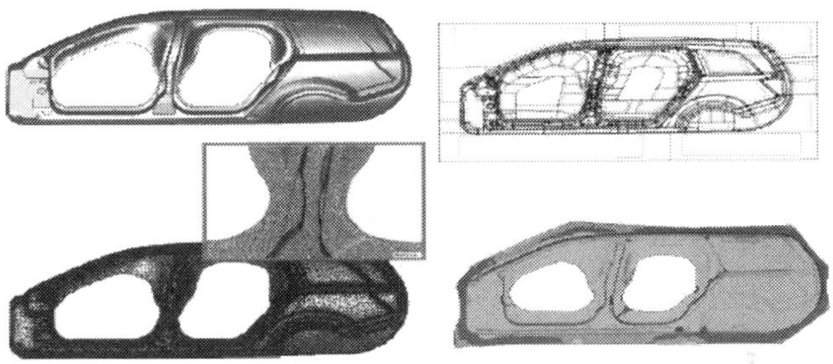

Figure. 1.7 (a) Actual geometry of an automotive panel. (b) CAD geometrical description by NURBS patches. (c) Finite element mesh of 3-noded shell triangles discretizing the panel geometry. (d) FEM numerical results of the structural analysis showing the equivalent strain distribution. Images by courtesy of Quantech ATZ S.A., www.quantech.es

Figure. 1.8 FEM analysis of the Agbar tower (Barcelona). Actual structure and discretization into shell and 3D beam elements. Deformed mesh (amplified) under wind load. Images are courtesy of Compass Ingenieria y Sistemas SA, www.compassis.com and Robert Brufau i Associats, S.A. www.robertbrufau.com.

CHAPTER 2

MATHEMATIC MODEL

MATRIX

Matrix concepts were first introduced in 1858 by Arthur Carley, a structural e3ngineers. In mathematics, a matrix is a rectangular array of numbers, symbols, or expressions, arranged in rows and columns. For example:

$$\begin{bmatrix} 1 & 2 & 3 \\ 4 & 5 & 6 \end{bmatrix}.$$

While structural engineering includes design and analysis problems. To solve a system of linear equations , Matrix become especially useful. Since the tedious algebraic operation become inefficient as the number of equations inefficient as the number of equations increases. For example, a typical system equations is shown below:

$$a_{11}x_1 + a_{12}x_2 + a_{13}x_3 + \cdots a_{1n}x_n = b_1$$
$$a_{21}x_1 + a_{22}x_2 + a_{23}x_3 + \cdots a_{2n}x_n = b_2$$
$$a_{31}x_1 + a_{32}x_2 + a_{33}x_3 + \cdots a_{3n}x_n = b_3$$
$$\vdots$$
$$a_{n1}x_1 + a_{n2}x_2 + a_{n3}x_3 + \cdots a_{nn}x_n = b_n$$

In matrix notation , expression (2-1) is rewritten as:

$$\begin{bmatrix} a_{11} & a_{12} & a_{13} & \cdots & a_{1n} \\ a_{21} & a_{22} & a_{23} & \cdots & a_{2n} \\ a_{31} & a_{32} & a_{33} & \cdots & a_{3n} \\ \vdots & & & & \\ a_{n1} & a_{n2} & a_{n3} & \cdots & a_{nn} \end{bmatrix} \begin{Bmatrix} x_1 \\ x_2 \\ x_3 \\ \vdots \\ x_n \end{Bmatrix} = \begin{Bmatrix} b_1 \\ b_2 \\ b_3 \\ \vdots \\ b_n \end{Bmatrix}$$

Expression can be also denoted, using capital letters to represent the matrix quantities, as

$$[A][X]=[B]$$

Where is the coefficient matrix, is the unknown matrix, and is the constant matrix.

To emphasize the value of the matrix equation, it is sufficient to re-arrange it to obtain the unknowns.

$$[X]=[A]^{-1}[B]$$

A matrix is made of elements. Each element has a numerical value as well as position coordinates which are described by their subscripts. The first subscript defines the row number and the second subscript defines the column number of its location. An element is at the intersection of the fifth row with the ninth column. It is convenient to use this notation because it facilitates some of the operations with matrices. A matrix having a single row is called a row matrix. The above discussion leads to the fact that matrices are rectangular arrays defined by the number of existing columns and rows. The elements of a matrix can be numbers, mathematical expressions, or alphanumerics. The matrices can be classified according to the combination of the individual values of the elements. Some matrices developed by this process have particular properties. To use them more efficiently it is helpful to be acquainted with them.

Concepts

Null Matrix:

A matrix whose elements are all zero.

$$[N]=[0]$$

The product of any matrix with null matrix is a null matrix.

Square Matrix:

A matrix having as many columns as rows. It is of a special type since it has a numerical value known as the determinant of the square matrix.

Identity Matrix:

A square matrix which has all zero elements except ones at its diagonal. It is also called a unit matrix. The product of any compatible matrix with an identity matrix is the initial matrix; .

$$\mathrm{I}=\begin{bmatrix}1 & 0 & 0\\ 0 & 1 & 0\\ 0 & 0 & 1\end{bmatrix}$$

$$[A]=[I]$$

$$if\ a_{ij}=1\ for\ \mathrm{i}=\mathrm{j}$$

$$a_{ij}=0\ \ for\, i\neq j$$

Symmetric Matrix:

A matrix whose elements are symmetric with respect to the diagonal. A matrix is symmetric if for all .

Skew-symmetric Matrix:

A matrix which has a negative symmetry with respect to the diagonal. A matrix is skew-symmetric if for all .

Diagonal Matrix:

A matrix whose elements are all zero except those on its diagonal. A unit matrix is a special type of a diagonal matrix. A matrix is a diagonal matrix if for all .

Triangular Matrix:

A matrix whose elements above or below the diagonal are all zero. There are two types of triangular matrices-upper and lower.

Column Matrix:

A matrix having its elements arranged in a single column form.

Row Matrix:

A matrix having its elements arranged in a single row form.

To Transpose a Matrix:

To interchange respectively the rows and columns of a matrix. The transpose of a product of two given matrices is the same as the product of the transpose of the second by the first matrix.

$$[A \cdot B]^T = [B]^T \cdot [A]^T$$

To Invert a Matrix:

To generate a new matrix using well-defined operations such that when it is multiplied with the original matrix it yields an identity matrix. The inversion operations are defined later in the chapter.

Determinant of a Matric:

A numerical value associated with each square matrix. The general expression which defines the calculation of the determinant of a matrix is given as:

$$|A| = \sum_{i=1}^{n} a_{ij} \cdot (cof\ a_{ij}) = \sum_{j=1}^{n} a_{ij} \cdot (cof\ a_{ij})$$

Singular Matrix:

A square matrix whose determinant is zero.

Orthogonal Matrix:

A matrix whose transpose and inverse are equal. A matrix whose transpose and inverse are equal. A matrix is an orthogonal matrix if .

Operation

In finite element method, the familiarity with matrix operations such as addition, multiplication, inversion, and equality is necessary.

Matrix Equality:

Two matrices are said to be equal if they have the same dimensions and the corresponding elements are identical. Two matrices have the same dimensions if they have the same number of rows and columns. is equal to if is equal to for all values of the subscripts .

Matrix Addition and Subtraction:

To add two or more matrices, they have to be compatible for addition. The compatibility for addition requires that the matrices have the same dimensions.

$$[A]+[B]+[C]=[D]$$

$$(a_{ij})+(b_{ij})+(c_{ij})=(d_{ij})$$

The associative and commutative laws hold true for matrix addition and subtraction.

Multiplication of Matrices:

To multiply two matrices the necessary condition is that the number of columns of the first must be equal to the number of rows of the second matrix.

$$[A]_{m\times n}\cdot[B]_{p\times r}=[C]_{m\times r}$$

To perform the above operation, n must be equal to p. The mathematical process of matrix multiplication is described by the following expression.

$$c_{ij}=\sum_{k=1}^{p}a_{ik}b_{kj}$$

To explain this formula, let us present an example.

$$\begin{bmatrix}2 & 5\\ 9 & 7\end{bmatrix}\begin{bmatrix}1 & 0\\ 4 & 3\end{bmatrix}=\begin{bmatrix}(2\times 1)+(5\times 4) & (2\times 0)+(5\times 3)\\ (9\times 1)+(7\times 4) & (9\times 0)+(7\times 3)\end{bmatrix}=\begin{bmatrix}22 & 15\\ 37 & 21\end{bmatrix}$$

The associative and distributive laws hold true for matrix multiplication.

$$[A\cdot B]\cdot[C]=[A]\cdot[B\cdot C]$$

$$[A]\cdot[B+C]=[A]\cdot[B]+[A]\cdot[C]$$

The commutative law does not hold true for matrix multiplication:

$$[A]\cdot[B]\neq[B]\cdot[A]$$

Before discussing the matrix inversion, we need to define basic concepts which are used in inversion operation. They are the determinant and the adjoint of a matrix.

Determinant, Adjoint and Inversion

Determinant

The determinant of a square matrix is a numerical value connected with this particular matrix and computed according to a well-defined operation.

For a matrix , where n is equal or greater than 3, there exists a powerful method to determine the value of the determinant. This method is known as expansion by cofactors.

$$|A| = \sum_{i=1}^{n} a_{ij} \cdot (cof\ a_{ij}) = \sum_{j=1}^{n} a_{ij} \cdot (cof\ a_{ij})$$

The cofactor of an element of a matrix is defined as

$$(cof\ a_{ij}) = (-1)^{i+j} (minor\ a_{ij})$$

Where is the determinant of the matrix obtained by deleting the th row and th column of the given matrix. For illustration, consider the following example.

For a 2×2 matrix, the determinant is defined as follows:

$$|A| = \begin{vmatrix} a_{11} & a_{12} \\ a_{21} & a_{22} \end{vmatrix} = a_{11}a_{22} - a_{21}a_{12}$$

For a 3×3 matrix, the determinant can be computed with a cross multiplication procedure. This is good only for a 3×3 matrix.

$$|A| = \begin{vmatrix} a_{11} & a_{12} & a_{13} \\ a_{21} & a_{22} & a_{23} \\ a_{31} & a_{32} & a_{33} \end{vmatrix} = a_{11}a_{22}a_{33} + a_{31}a_{23}a_{12} + a_{32}a_{21}a_{13} - a_{31}a_{22}a_{13} - a_{31}a_{22}a_{13} - a_{32}a_{23}a_{11} - a_{33}a_{21}a_{12}$$

Another effective method of evaluating a determinant with numerical elements is by the "pivotal condensation" scheme. This approach allows the computation of the determinant of a square matrix by 2×2 subdeterminants.

$$|A| = \begin{bmatrix} a_{11} & a_{12} & \cdots & a_{1n} \\ a_{21} & a_{22} & \cdots & a_{2n} \\ \vdots & \vdots & & \vdots \\ a_{n1} & a_{n2} & \cdots & a_{nn} \end{bmatrix}$$

To facilitate the calculation of determinants, the following six rules will hold true:

a Interchanging two tows or two columns of a square matrix will alter only the sign of its determinant.

b Multiplying each element of a row or a column of a square matrix by a constant is the same as multiplying the matrix by the same constant.

c If two rows or columns of a square matrix are equal or multiple of each other, the value of the determinant is zero.

d If all the elements of a row or a column are zero, then the determinant is zero.

e The transpose of a square matrix will not change its determinant.

f The determinant of a square matrix will not change if a column or a row is multiplied by a constant and added to a column or a row.

Adjoint

The adjoint of a matrix [A] is a matrix which is the transpose of the cofactor matrix of [A]. In general notation, this is expressed as

$$\text{adj}[\mathrm{A}] = [cofA]^T$$

Inversion

There are several methods available for matrix inversion. We will explain and illustrate the relevant methods. One has to remember that only square matrices have inverses.

1 Inversion by Adjoint Method

This method is an elimination procedure without decreasing the number of equations in the system. A unit matrix of the same dimension as the original matrix is attached to the right of the original matrix. Performing the proper matrix row operations, the original matrix is reduced to a unit matrix; the same row operation are applied to the attached unit matrix, which becomes the inverse matrix.[(A|I)] will become [(I/B)] when $[B] = [A]^{-1}$.

Examples

The solution of many technical problems requires the analysis of a network system formed by different \elements “connected by their extremities or joints, and subjected to a set of \loads” which are usually external to the system. Examples of such systems, which we will call discrete systems, are common in structural engineering (pin-jointed bar structures, frames, grillages, etc.) and in many other different engineering problems, e.g.: hydraulic piping networks, electric networks, transport planning networks, production organization systems (PERT, etc.) amongst others. Figure 1.8 shows some of these discrete systems.

Discrete systems can be studied using matrix analysis procedures which have a very close resemblance to the FEM.

Basic concepts of matrix analysis of bar structures

Matrix analysis is the most popular technique for the solution of bar structures.

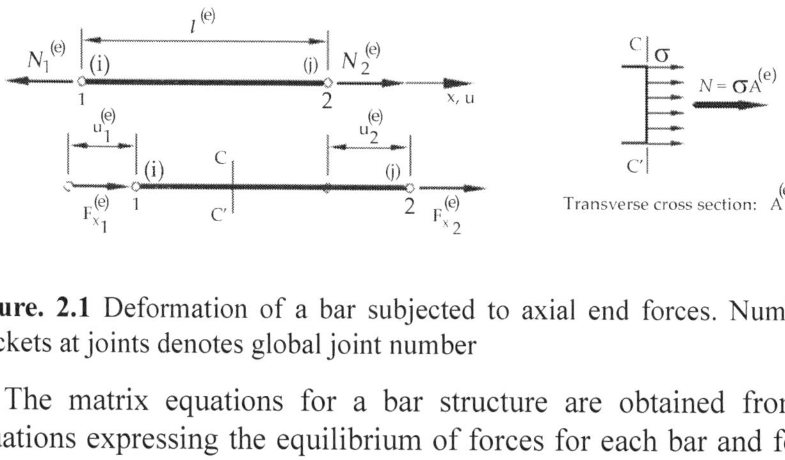

Figure. 2.1 Deformation of a bar subjected to axial end forces. Number in brackets at joints denotes global joint number

The matrix equations for a bar structure are obtained from the equations expressing the equilibrium of forces for each bar and for the structure as a whole. Let us consider an isolated bar, e, of length $l^{(e)}$ subjected to axial forces $F_{x_1}^{(e)}$ and $F_{x_2}^{(e)}$ acting at the beam joints (Figure 2.1). The x axis has the direction of the bar. Strength of Materials defines the strain at any point in the bar by the relative elongation, i.e.

$$\varepsilon = \frac{\Delta l^{(e)}}{l^{(e)}} = \frac{u_2^{(e)} - u_1^{(e)}}{l^{(e)}}$$

where $u_1^{(e)}$ and $u_2^{(e)}$ are the displacements of the joint points 1 and 2 in the x direction, respectively. In equations denotes values associated to an individual bar. Generally indexes 1 and 2 are local joint numbers for the bar and correspond to the actual global numbers i, j of the joints in the structure. Hence $u_1^{(e)} = u_i$ and $u_2^{(e)} = u_j$.

The axial stress ó is related to the strain å " by Hooke law as

$$ó = \text{E}å = \text{E}\frac{u_2^{(e)} - u_1^{(e)}}{l^{(e)}}$$

where E is the Young modulus of the material. The axial force N at each section is obtained by integrating the stress over the cross sectional area. The axial force N is transmitted to the adjacent bars through the joints. For homogeneous material we have

$$\aleph_2^{(e)} = A^{(e)}\sigma = (EA)^{(e)}\frac{u_2^{(e)} - u_1^{(e)}}{l^{(e)}} = \aleph_1^{(e)}$$

The force equilibrium equation for the bar of Figure 2.1 is simply

$$F_{x_1}^{(e)} + F_{x_2}^{(e)} = 0$$

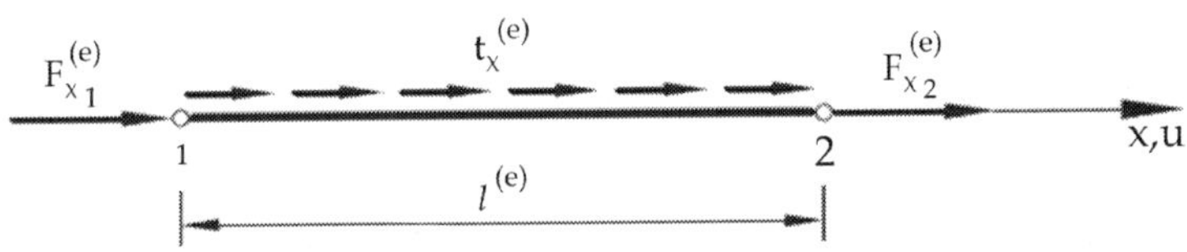

$$F_{x1}^{(e)} + \frac{(lt_x)^{(e)}}{2} = k^{(e)}(u_1^{(e)} - u_2^{(e)})$$

$$F_{x2}^{(e)} + \frac{(lt_x)^{(e)}}{2} = k^{(e)}(u_2^{(e)} - u_1^{(e)}), \quad k^{(e)} = \left(\frac{EA}{l}\right)^{(e)}$$

Figure. 2.2 Equilibrium equations for a bar subjected to axial joint forces and a uniformly distributed axial load $t_x^{(e)}$

A uniformly distributed external axial load of intensity $t_x^{(e)}$ can easily be taken into account by adding one half of the total external load to each axial force at the bar joints. The equilibrium equations now read (Figure 1.10).

$$q^{(e)} = \begin{Bmatrix} F_{x_1}^{(e)} \\ F_{x_2}^{(e)} \end{Bmatrix} = k^{(e)} \begin{bmatrix} 1 & -1 \\ -1 & 1 \end{bmatrix} \begin{Bmatrix} u_1^{(e)} \\ u_2^{(e)} \end{Bmatrix} - \frac{(lt_x)^{(e)}}{2} \begin{Bmatrix} 1 \\ 1 \end{Bmatrix} = K^{(e)} a^{(e)}$$

Where

$$f^{(e)} = \begin{Bmatrix} f_{x_1}^{(e)} \\ f_{x_2}^{(e)} \end{Bmatrix} = \frac{(lt_x)^{(e)}}{2} \begin{Bmatrix} 1 \\ 1 \end{Bmatrix}$$

Basic steps for matrix analysis of discrete systems

What we have seen this far leads us to conclude that the analysis of a discrete system (i.e. a bar structure) involves the following steps:

a) Definition of a network of discrete elements (bars) connected among themselves by joints adequately numbered. Each element e has known geometrical and mechanical properties. All these characteristics constitute the problem data and should be defined in the simplest possible way (preprocessing step).

b) Computation of the stiffness matrix K(e) and the joint force vector f(e) for each element of the system.

c) Assembly and solution of the resulting global matrix equilibrium equation (Ka = f) to compute the unknown parameters at each joint, i.e. the displacements for the bar system.

d) Computation of other relevant parameters for each element, i.e. the axial strain and the axial force, in terms of the joint parameters.

The results of the analysis should be presented in graphical form to facilitate the assessment of the system's performance (post-processing step).

Example:

Compute the displacements and axial forces in the three-bar structure of Figure 2.3 subjected to an horizontal force P acting at its right hand end.

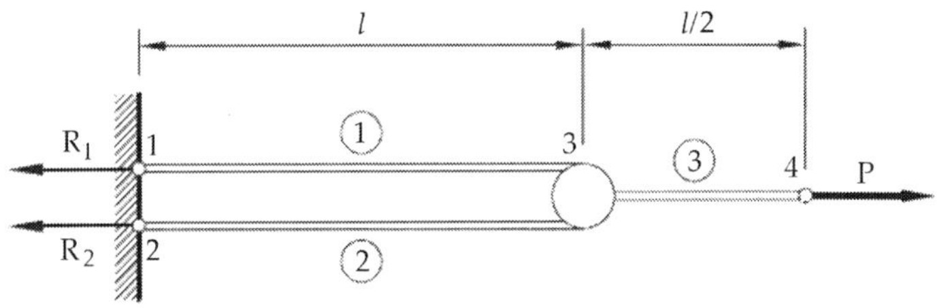

E and A are the same for each bar

Figure 2.3 Analysis of a simple three-bar structure under an axial load

Solution

The equilibrium equations for each joint are

$$\text{Bar1}\begin{Bmatrix} F_{x_1}^{(1)} \\ F_{x_2}^{(1)} \end{Bmatrix} = k^{(1)} \begin{bmatrix} 1 & -1 \\ -1 & 1 \end{bmatrix} \begin{Bmatrix} u_1^{(1)} \\ u_2^{(1)} \end{Bmatrix}$$

$$\text{Bar2}\begin{Bmatrix} F_{x_1}^{(2)} \\ F_{x_2}^{(2)} \end{Bmatrix} = k^{(2)} \begin{bmatrix} 1 & -1 \\ -1 & 1 \end{bmatrix} \begin{Bmatrix} u_1^{(2)} \\ u_2^{(2)} \end{Bmatrix}$$

$$\text{Bar3}\begin{Bmatrix} F_{x_1}^{(3)} \\ F_{x_2}^{(3)} \end{Bmatrix} = k^{(3)} \begin{bmatrix} 1 & -1 \\ -1 & 1 \end{bmatrix} \begin{Bmatrix} u_1^{(3)} \\ u_2^{(3)} \end{Bmatrix}$$

With $k^{(1)} = k^{(2)} = \dfrac{EA}{l}$ and $k^{(3)} = \dfrac{2EA}{l}$

The compatibility equations between local and global displacements are:

$$u_1^{(1)} = u_1;\quad u_2^{(1)} = u_3\ ;\quad u_1^{(2)} = u_2$$

$$u_2^{(2)} = u_3;\quad u_1^{(3)} = u_3\ ;\quad u_2^{(3)} = u_4$$

Applying the assembly equation to each of the four joints we have

$$\text{joint1}: \sum_{e=1}^{3} F_{x_i}^{(1)} = -R_1 \quad , \qquad joint\,2: \sum_{e=1}^{3} F_{x_i}^{(1)} = -R_2$$

$$\text{joint3}: \sum_{e=1}^{3} F_{x_i}^{(1)} = 0 \quad , \qquad joint\,4: \sum_{e=1}^{3} F_{x_i}^{(1)} = P$$

Substituting the values of $F_{x_i}^{(e)}$ from the bar equilibrium equations gives

$$\text{joint1}: k^{(1)}\left(u_1^{(1)} - u_2^{(1)}\right) = -R_1$$

$$joint\,2: k^{(1)}\left(u_1^{(2)} - u_2^{(2)}\right) = -R_2$$

$$\text{joint3}: k^{(1)}\left(-u_1^{(1)} + u_2^{(1)}\right) + k^{(2)}\left(-u_1^{(2)} + u_2^{(2)}\right) + k^{(3)}\left(u_1^{(3)} + u_2^{(3)}\right) = 0$$

$$joint\,4: k^{(3)}\left(-u_1^{(3)} + u_2^{(1)}\right) = P$$

Above equations can be written in matrix form using the displacement compatibility conditions as

$$\begin{bmatrix} k^{(1)} & 0 & -k^{(1)} & 0 \\ 0 & k^{(2)} & -k^{(2)} & 0 \\ -k^{(1)} & -k^{(2)} & \left(k^{(1)} + k^{(2)} + k^{(3)}\right) & -k^{(3)} \\ 0 & 0 & -k^{(3)} & k^{(3)} \end{bmatrix} \begin{Bmatrix} u_1 \\ u_2 \\ u_3 \\ u_4 \end{Bmatrix} = \begin{Bmatrix} -R_1 \\ -R_2 \\ 0 \\ P \end{Bmatrix}$$

Note that an external point load acting at node j can be placed directly in the jth position of the global joint force vector f.

Substituting the value of k^e for each bar and imposing the boundary conditions $u_1 = u_2 = 0$, the previous system can be solved to give

$$u_3 = \frac{Pl}{2EA}$$

$$u_4 = \frac{Pl}{EA}$$

$$R_1 = R_2 = \frac{P}{2}$$

The axial forces in each bar are finally obtained as

$$\text{Bar1}: \aleph^{(1)} = \frac{EA}{l}\left(u_3 - u_1\right) = \frac{P}{2}$$

$$\text{Bar2}: \aleph^{(2)} = \frac{EA}{l}\left(u_3 - u_2\right) = \frac{P}{2}$$

$$\text{Bar3}: \aleph^{(3)} = \frac{2EA}{l}\left(u_4 - u_3\right) = P$$

The joint axial force for each bar is computed

$$\aleph_2^{(1)} = -\aleph_1^{(1)} = \frac{P}{2}$$

$$\aleph_2^{(2)} = -\aleph_1^{(2)} = \frac{P}{2}$$

$$\aleph_2^{(3)} = -\aleph_1^{(3)} = P$$

CHAPTER 3

COMMERCIAL SOFTWARE OF FINITE ELEMENT METHOD

There are numerous companies selling FEM software. Many of the smaller companies have either closed or have been absorbed by larger companies.

Some of this began with the purchase of Fluid Dynamics International (developers of FIDAP, a penalty approach FEM) by Fluent, Inc., which sold FLUENT, a popular CFD finite volume code. Fluent later obtained NEKTON, a spectral element code developed at MIT. Fluent, Inc. was subsequently purchased by ANSYS, Inc., who had also purchased FLOTRAN (an equal-order FEM code). COMCO, Inc., who had originally developed an hp-adaptive flow code in UT-Austin, and CENTRICS, a high-end finite element CFD code first developed at Stanford, both closed. The choices are numerous, and sometimes confusing for the buyer interested in obtaining a good code, especially one that is FEM based. COMSOL is a multiphysics FEM code that is widely used for its ability to couple various physics problems, e.g., fluid-structural-heat transfer, together into one overall solution strategy. This is a very easy program to use, reasonably priced and runs on PCs, with follow-on support from

COMSOL, Inc. This chapter will introduce three most popular softwares : ABAQUS, ANSYS and COMSOL.

ABAQUS

Introduction

Abaqus is used in the automotive, aerospace, and industrial product industries. The product is popular with academic and research institutions due to the wide material modeling capability, and the program's ability to be customized. Abaqus also provides a good collection of multiphysics capabilities, such as coupled acoustic-structural, piezoelectric, and structural-pore capabilities, making it attractive for production-level simulations where multiple fields need to be coupled.

It uses finite element methods to find complete solutions for simple to complex linear and nonlinear engineering problems, the address non-linear physical behavior makes the software package has an extensive range of material models such as elastomeric material capabilities.

The Abaqus finte element system includs:

- Abaqus/Standard, a general-purpose finite element program;
- Abaqus/Explicit, an explicit dynamics finite element program;
- Abaqus/CAE, and interactive environment used to create finite element models, submit Abaqus analysis, monitor and diagnose jobs, and evaluate results;
- Abaqus/Viewer, a subset of Abaqus/CAD that contains only the postprocessing capabilities of the Visualization module;
- Abaqus/Aqua option works with Abaqus/Standard and Abaqus/ Explicit. Abaqus/Aqua contains optional features that are specifically designed for the analysis of beam-like structures installed underwater and subject to loading by water current and wave action;
- Abaqus/Design option enables you to perform design sensitivity analysis(DSA). The Abaqus/Design option works with Abaqus/Standard;
- Abaqus/AMS is an optional eigensolver that works within

Abaqus/Standard providing very fast solution of large symmetric eigenvalue problems;

- Abaqus co-simulation technique provides two applications, available as esparate add-on capabilities, for coupling between Abaqus the third-party analysis programs;
- Abaqus/Foundation is an optional subset of Abaqus/Standard that provides more cost-efficient access to the linear static and dynamic analysis functionality in Abaqus/Standard.

Example

This example is about welded lug which is fixed to one end, while the other has a hold. You need to plug a bolt into the hole if you want to use it. What is deflection of it if use 30 KN load to the plugged bolt. Commercial software Abaqus has a standard mode, it is perfect match its static situation. Before use it, some assumption need to be decided as shown below,

Distributed load worked on the down part of the lug hole, which is close to the rod. It is set in the model rather other complex physical reaction, it was shown in Figure 4.1.

As mentioned above, all other load which is not distributed but various value was uniformed to the set pressure.

The set load which was in pressure form was $50\ \text{MPa}$, which is got from $30\ \text{KN}/\left(2\times 0.015\times 0.02m^2\right)$

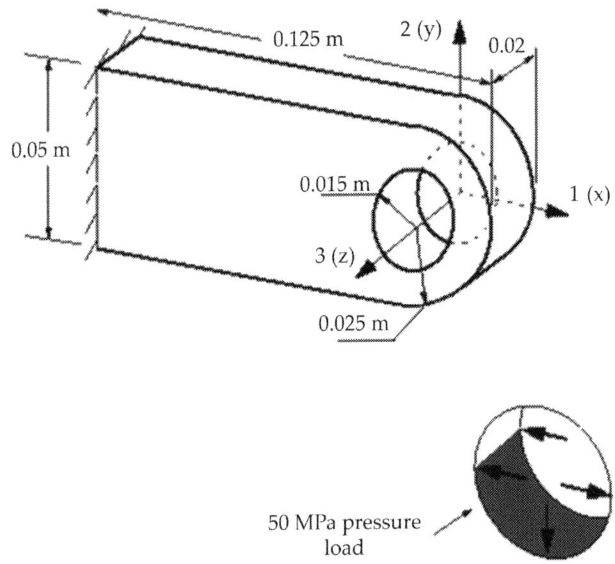

Figure 4.1 Geometry of connecting lug.

Modeling

3D modeling:

1 Give a name to the part we created, we called it "Lug" in the shown box. Other setting of the part we created is set to be default value. So it should be a 3D soild body.For the size which need to write as approximation one should be two times large as the largest value. We will input the 0.25 here.

To get the figure shown in Figure 4.2, the bellow operation need to be done.

a) w a rectangle by the drawing tool, here is the rectangle tool. Then use the constrain tool to constrain the rectangle to be the ideal shape, the applying the dimensional tool to constrain the size of the shape. Here it is 0.1×0.05 m, the geometry is shown in Figure 4.2. The two tool buttons were as and .

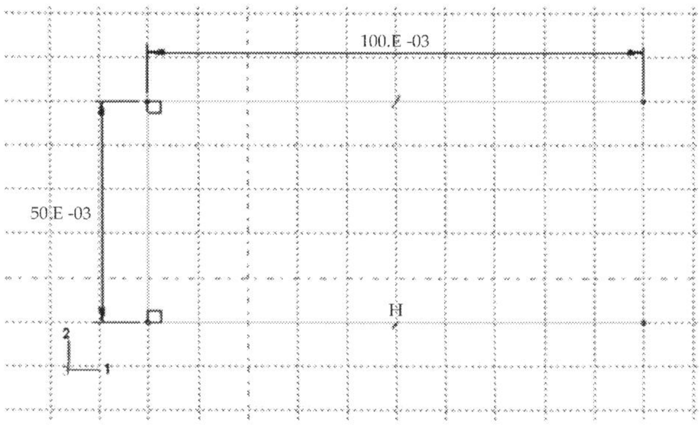

Figure 4.2 Sketch of the rectangle shape .

Note: This figure has already added size and constrain conditions to get the required geometry of the model. This operations can be done from the sketch section or clicking the related buttons in the menu.

b) get the lug shape, an arc shaped end also need to added on. As shown in Figure 4.3, adding the arc shape by clicking the thru three points button. Select the two end points of the rectangle to draw the arc. Deleting the arc side's line. Then constrain the arc and the line to be tangent with each other. Then the required shape has obtained.

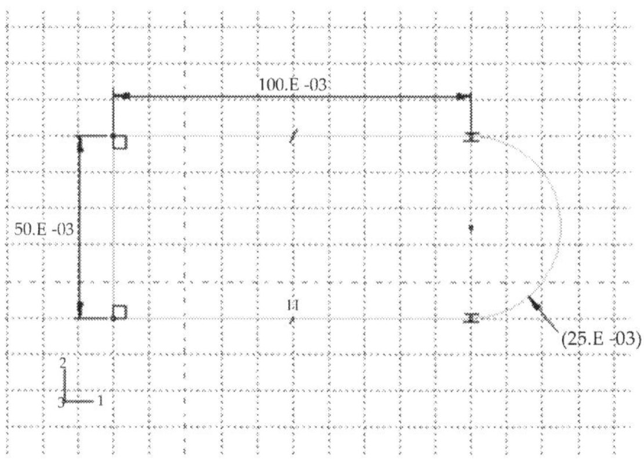

Figure 4.3 adding the arc.

Setting Material property

Besides the geometry part of the part of the model. Another thing we need to set is the material property. For material property, it should be setting each constant to define a particular property for example elastic property and plastic property. Here we define the model's material property by providing the Young's modulus and Possion ratio separately with 200 GPa and0.3 respectively.

Besides the material property, in Abaqus, the section property also need to be assigned.

Assembly

The assembly here is different with the assembly we will introduced in the model. Here it just means the assembly of the geometry parts. Adding all exist parts to the empty assembly model.

Defining Steps and Specifying Output Requests

You will now define the analysis steps. Since interactions, loads, and boundary conditions can be step dependent, analysis steps must be defined before these can be specified. For this simulation you will define a single static, general step. In addition, you will specify output requests for your analysis. These requests will include output to the output database (.odb) file.

Set Steps

For a complete simulation, each steps need to be clarified by defining the catalog of the mechanics (static or dynamic), interactions and BC. After setting the steps, the output request need to be set also. The output will be generated by "obd" form. The output need to be specified the exact results that you want to see by the visulation, i.e, Stress or displacement.

BC

BC is short for boundary condtion. The boundary condition need to be clarified before processing the model. The boundary condition can be the constrains in each dimension on the model's geometry. In this example, the boundary condition is shown by Figure 4.4 with one surface fixed. After specify the boundary condition, the loads condition should also be specified. In here, it is 50 MPa around the end hole.

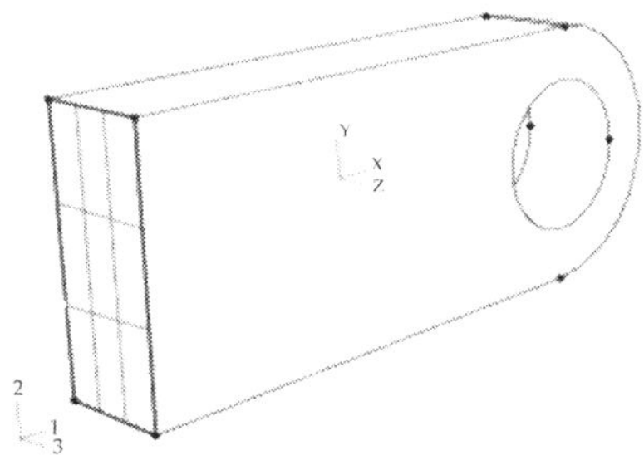

Figure 4.4 Built-in end of the connecting lug.

Mesh

Meshing is one of the most important process during the FEM analysis. The meshing quality will greatly affect the analysis results. For different model geometry, there are different suite mesh type. Mesh number is another thing the user need to consider besides the type.

For example, the lug has meshed by hexahedral elements. The seed size is 0.007 with first order elements, another option is using seed size 0.01 for second order elements. The meshing work can be seen from Figure 4.5.

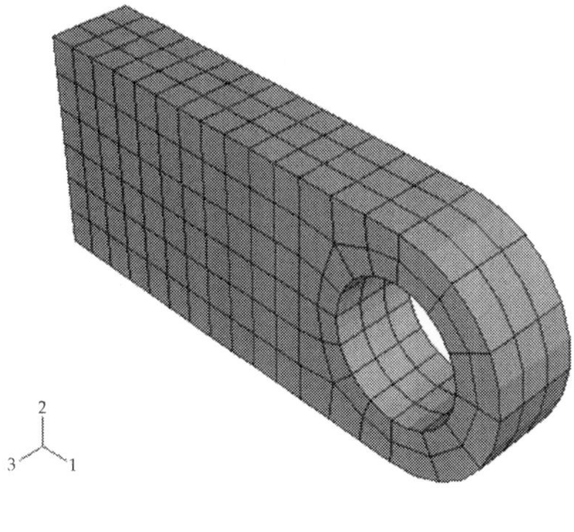

Figure 4.5 Suggested mesh of C3D20R elements for the connecting lug model.

Running

This is what Abaqus did during computation. For users, what you need to do is just "create a job", then "running the job", then the abaqus will use all the parameters you have already set to summit the job to the backstage for analysis. It replaces the complex calculation.

Post-processing

Post-processing can deal with the deformed shapes, as shown in Figure 4.6, or the changed view shown in Figure 4.7.

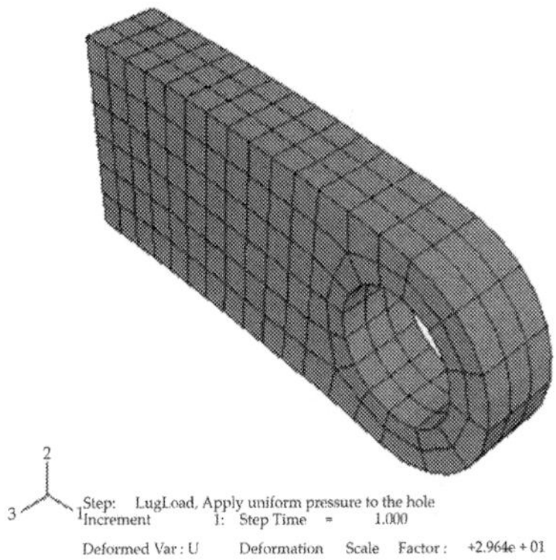

Figure 4.6 Deformed model shape of connecting lug (shaded).

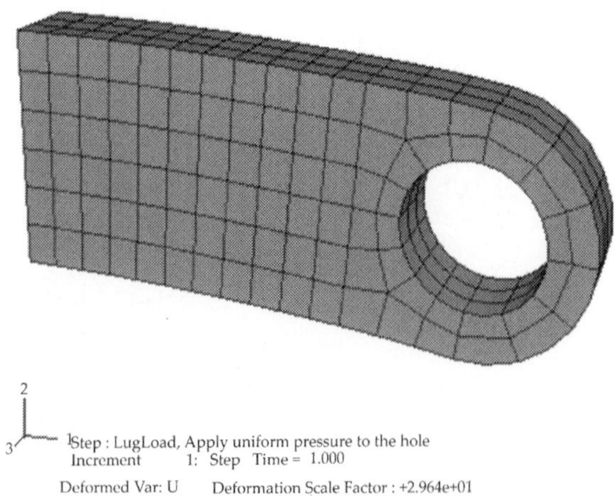

Figure 4.7 Deformed model shape viewed from specified viewpoint.

You can also observe the whole picture without the mesh, as shown in Figure 4.8.

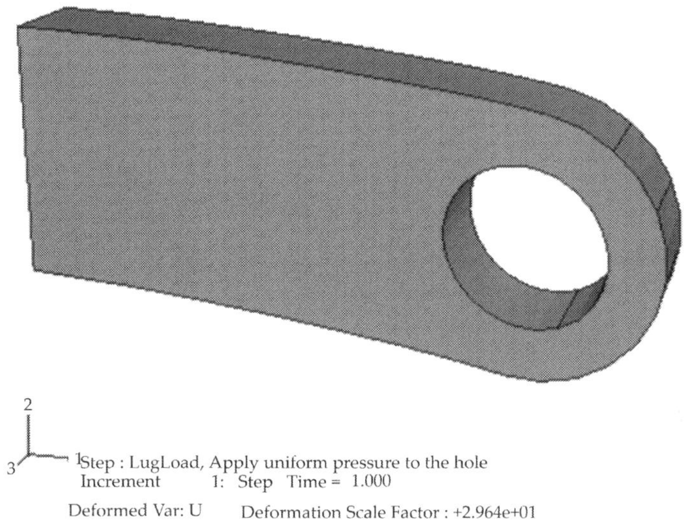

Figure 4.8 Deformed shape with only feature edges visible.

Or just to shown the edge without the solid body as shown in Figure 4.9, or hidden several lines as shown in 4.10.

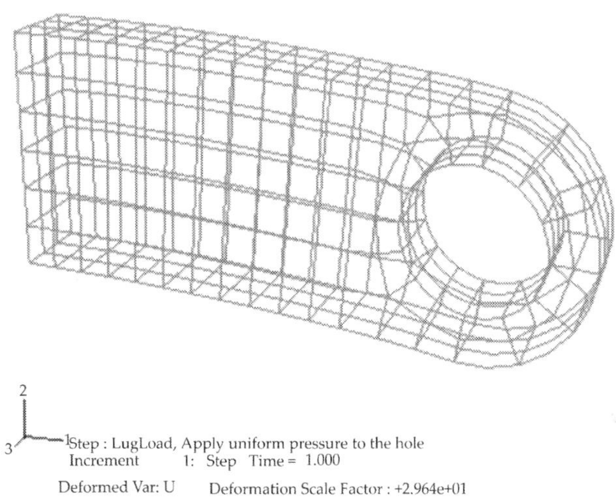

Figure 4.9 Wireframe plot.

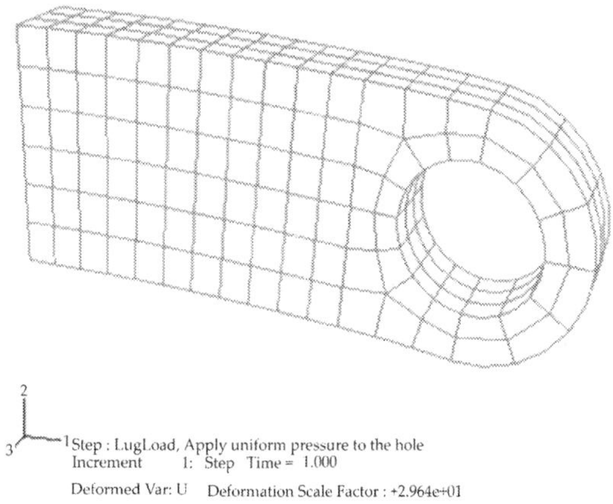

Figure 4.10 Hidden line plot.

Figure 4.11 and Figure 4.12 shows the plot of the stress which set by default set and customized set.

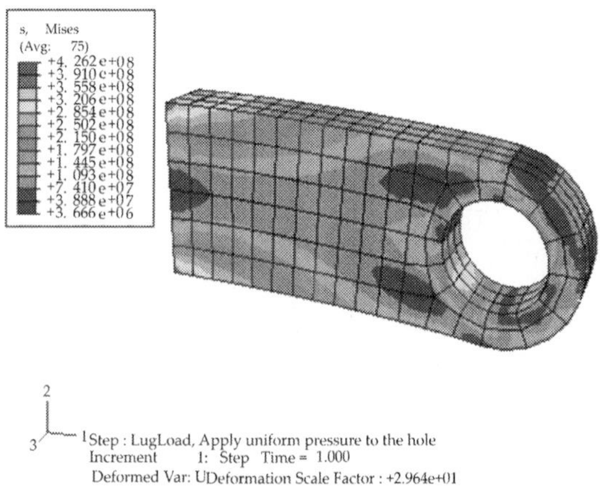

Figure 4.11 Filled contour plot of Mises stress.

By Abaqus, you can also set the stress plot yourself by setting the related values. If you use the default set, this software will set the range of the largest and smallest value of the stress by the calculation results and then divide them into twelve parts. However you can set these two values yourself, even the divided parts number. As shown in Figure 4.11.

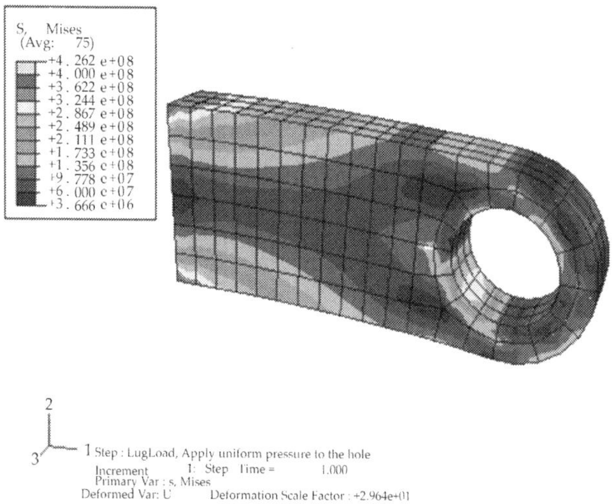

Figure 4.11 Customized plot of Mises stress.

ANSYS

Introduction

ANSYS Mechanical software is a comprehensive FEA analysis (finite element) tool for structural analysis, including linear, nonlinear and dynamic studies. The engineering simulation product provides a complete set of elements behavior, material models and equation solvers for a wide range of mechanical design problems.

Example

A steel cylinder with inner radius 5 inches and outer radius 10 inches is 40 inches long and has spherical end caps. The interior of the cylinder is kept at 75 deg F, and heat is lost on the exterior by convection to a fluid whose temperature is 40 deg F. The convection coefficient is 0.56 BTU/hr-sq.in-F. Calculate the stresses in the cylinder caused by the temperature distribution.

The problem is solved in two steps. First, the geometry is created, the preference set to 'thermal', and the heat transfer problem is modeled and solved. The results of the heat transfer analysis are saved in a file 'jobname.RTH' (Results THermal analysis) when you issue a save jobname.db command.

Next the heat transfer boundary conditions and loads are removed from the mesh, the preference is changed to 'structural', the element type is changed from 'thermal' to 'structural', and the temperatures saved in 'jobname.RTH' are recalled and applied as loads.

- Start ANSYS and assign a job name to the project. Run Interactive -> set working directory and jobname.
- Preferences -> Thermal will show -> OK
- A quadrant of a section through the cylinder is created using ANSYS area creation tools.

- Preprocessor -> Element Type -> Add/Edit/Delete -> Add -> Solid 8 node 77 -> OK ->Options -> K3 Axisymmetric -> OK
- Preprocessor -> Material Props -> Isotropic -> Material Number 1 -> OK

EX = 3.E7 (psi)

DENS = 7.36E-4 (lb sec^2/in^4)

ALPHAX = 6.5E-6

PRXY = 0.3

KXX = 0.69 (BTU/hr-in-F)

- Mesh the area using methods discussed in previous examples.
- Preprocessor -> Loads -> Apply -> Temperatures -> Nodes
- Select the nodes on the interior and set the temperature to 75.
- Preprocessor -> Loads -> Apply -> Convection -> Lines
- Select the lines defining the outer surface and set the coefficient to 0.56 and the fluid temp to 40.
- Preprocessor -> Loads -> Apply -> Heat Flux -> Lines

Select the vertical and horizontal lines of symmetry and set the heat flux to zero.

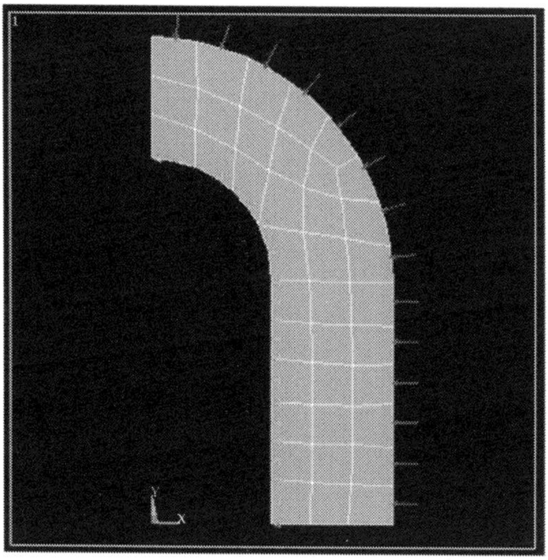

- Solution -> Solve current LS
- General Postprocessor -> Plot Results -> Nodal Solution -> Temperature

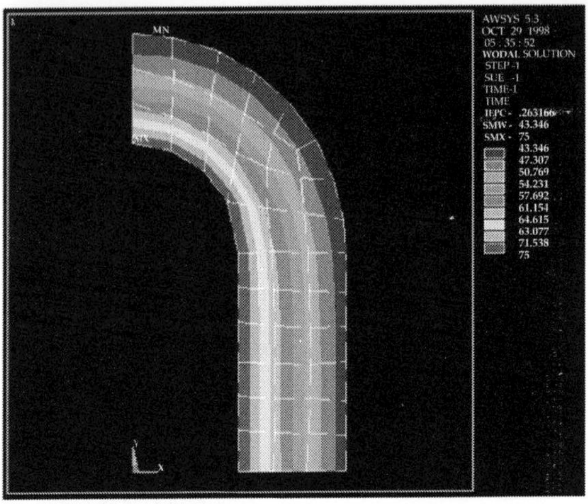

- File -> Save Jobname.db
- Preprocessor -> Loads -> Delete -> Delete All -> Delete All Opts.
- Preferences -> Structural will show, Thermal will NOT show.
- Preprocessor -> Element Type -> Switch Element Type -> OK (This changes the element to structural)
- Preprocessor -> Loads -> Apply -> Displacements -> Nodes
- (Fix nodes on vertical and horizontal lines of symmetry from crossing the lines of symmetry.)
- Preprocessor -> Loads -> Apply -> Temperature -> From Thermal Analysis

Select Jobname.RTH (If it isn't present, look for the default 'file. RTH' in the root directory)

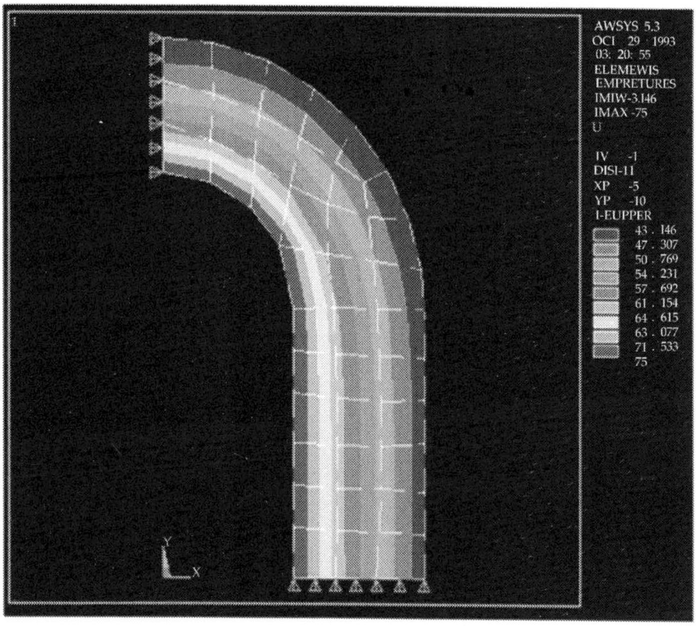

Solution -> Solve Current LS

General Postprocessor -> Plot Results -> Element Solution - von Mises Stress

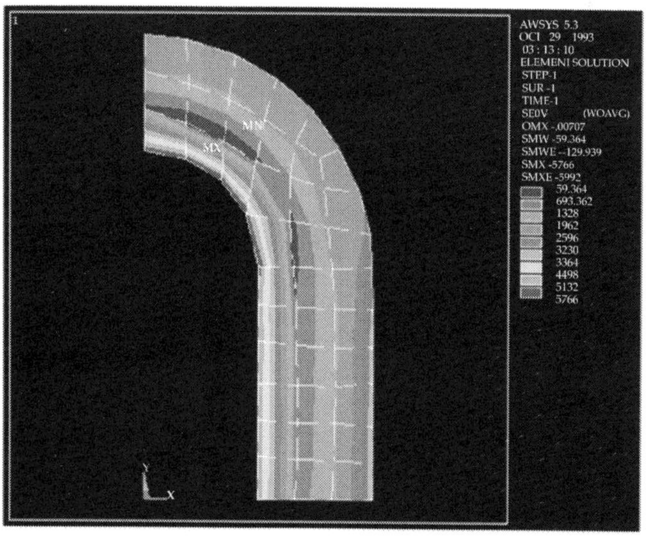

The von Mises stress is seen to be a maximum in the end cap on the interior of the cylinder and would govern a yield-based design decision.

The problem described in the axisymmetric analysis example will now be solved using 3 dimensional modeling and analysis. (Of course, three-D modeling is not necessary for this problem, but it serves as an example for us, and we have an exact solution available for comparison.)

To restate the problem, consider the problem of finding the stresses in a thick open-ended cylinder with an internal pressure (such as a pipe discharging to the atmosphere). The steel cylinder below has an inner radius of 5 inches and an outer radius of 11 inches.

A cross section is shown below

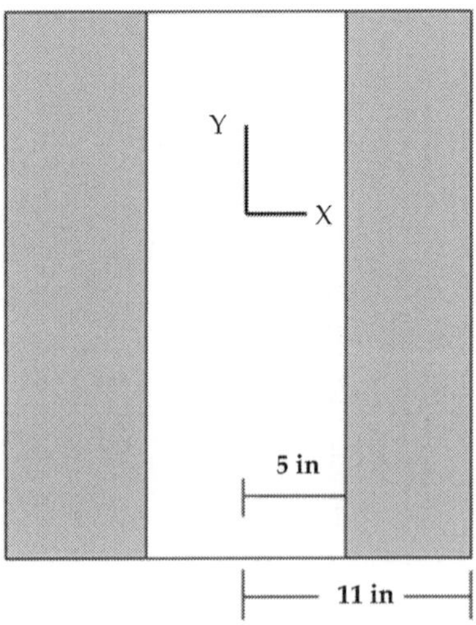

In both drawings the length is arbitrary and represents a segment of a long, open-ended cylinder. The Y axis is the axis of symmetry. The cylinder can be generated by revolving a rectangle 6 inches wide and of arbitrary height 360 degrees about the Y axis.

For the 3D analysis we will make use of symmetry and analyze one quadrant of the cylinder. If Y is the axis of revolution of the cylinder, both the X-Y and Y-Z planes are planes of symmetry.

Since the height of the segment considered is arbitrary, we will use a segment 1 inch in height for the finite element model. The geometry is shown below.

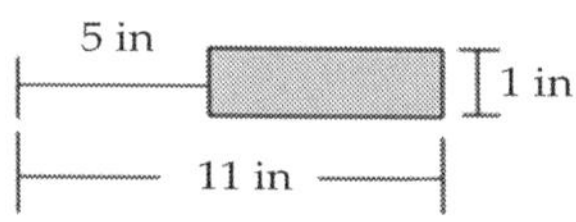

The geometry is created using ANSYS geometry commands. (Alternatively the geometry could be created in a solid modeler such as Pro/ENGINEER, CATIA, SolidWorks, etc and imported into ANSYS for FEM analysis. See ANSYS Example 8.B)

The ANSYS solution to this problem is given next for an internal pressure of 1000 psi.

1. Start ANSYS

Use a tetrahedron element for the FEM modeling.

2. Preprocessor -> Element Type -> Add/Edit/Delete -> Add -> select Tet 10node 92 -> OK -> Close

Enter material property data for steel.

3. Preprocessor -> Material Props -> Material Models . . . -> Structural -> Linear -> Elastic -> Isotropic ->Enter EX = 3.e7 and PRXY = 0.3 -> OK -> Close.

Create geometry for rectangle 1 inch by 6 inches starting 5 inches from Y axis. This rectangular area will be revolved 90 degrees about the Y axis to produce the desired volume.

4. Preprocessor -> Create -> Areas -> Rectangle ->By 2 Corners

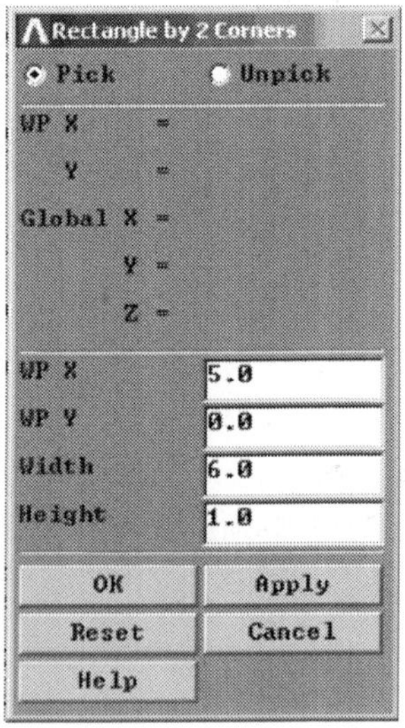

Define two key points on the Y axis for revolving purposes.

5. In the gray box 'Pick a menu item or enter an ANSYS command' enter.

K, 10, 0.0, 0.0, 0.0 (Return) Defines key point number 10 at (0, 0, 0)

K, 11, 0.0, 5.0, 0.0 (Return) Defines key point number 11 at (0, 5, 0)

(Key points 1,2,3,4 are already used by ANSYS to define the rectangle, so we used numbers 10 and 11.)

Revolve the area

6. Preprocessor -> Operate -> Extrude -> Areas -> About Axis

Pick the area then pick the two key points on the Y axis.

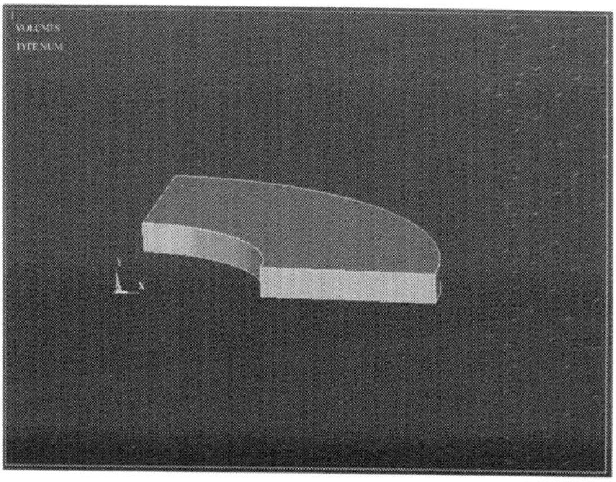

Now create a mesh of tetrahedral elements using the default settings.

5. Preprocessor -> Mesh -> Volumes -> Free (Pick the solid we just created.)

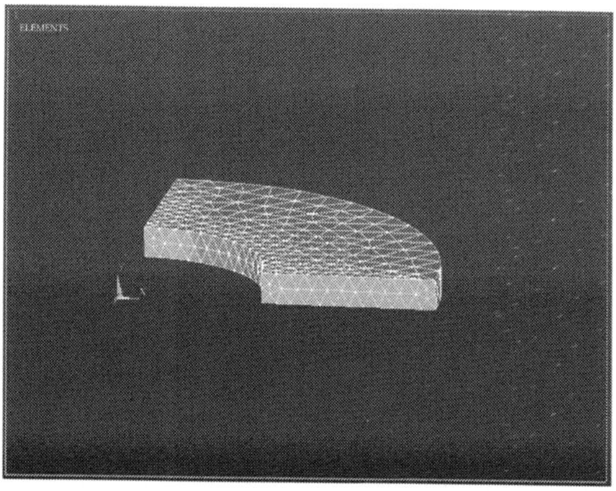

Define loads and boundary conditions. Apply an internal pressure of 1000 psi and displacement constraints that prevent points moving across planes of symmetry and also restrain rigid body movement in the Y direction.

6. Preprocessor -> Loads ->Apply ->Displacement -> On Areas Pick the radial surface parallel to the X-Y plane and set uz = 0 . Pick the radial surface parallel to the Z-Y plane and set ux = 0 . Pick the bottom (or top

surface) and set uy = 0 . This last constraint simply prevents rigid body motion in the Y direction.

Note: Use Cntl + Right Mouse to rotate the solid so as to get a better view of the area you want to pick. It may also help to activate the area numbers and plot them. Then pick the area number of interest, eg,

PlotCntls -> Numbering . . . turn area numbers ON

Plot -> Areas.

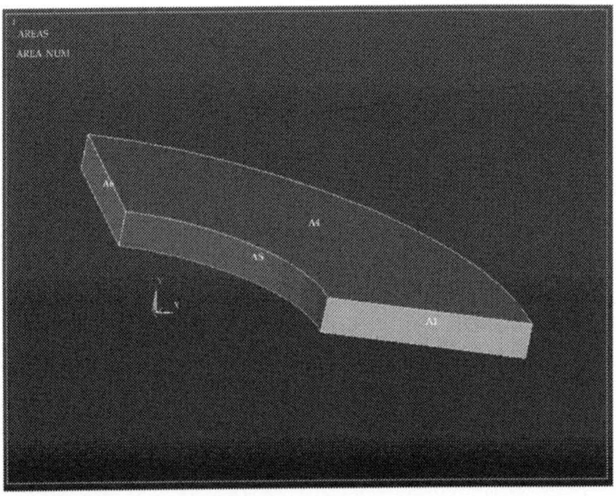

Preprocessor -> Loads ->Apply ->Pressure -> On Areas Pick the inner area of the cylinder). Enter a pressure of 1000.

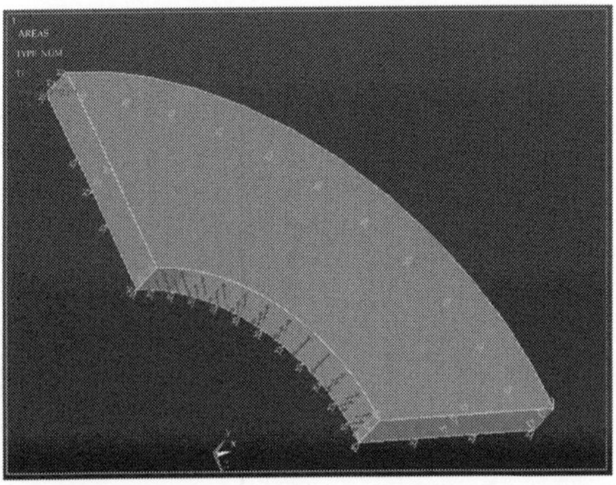

For large 3D problems it is often much faster to use the optional Preconditioned Conjugate Gradient method for solving the equations.

7. Solution -> Sol'n Control . . . -> Sol'n Options -> Pre-Condition CG -> OK

Solution -> Solve -> Current LS ->OK

Check the deformed shape to see if it's reasonable. (The dotted line is the undeformed shape.)

8. General Post Processor -> Plot Results -> Deformed Shape . . . ->Def +undeformed -> OK

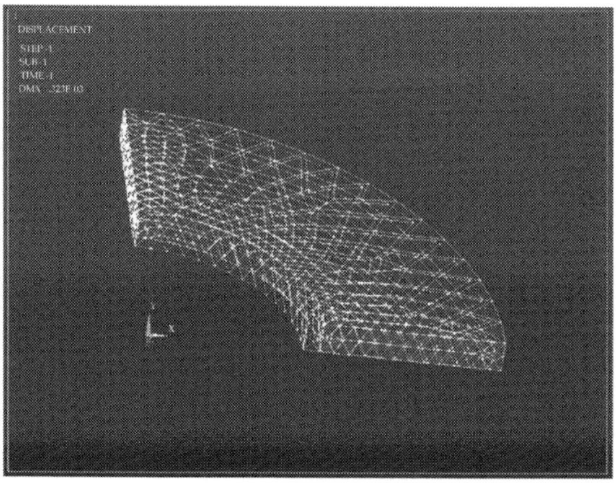

Examine the stresses ----------

9. General Postprocessor -> Plot Results -> Element Solu . . . (Pick Sx).

SX -------------------------------------

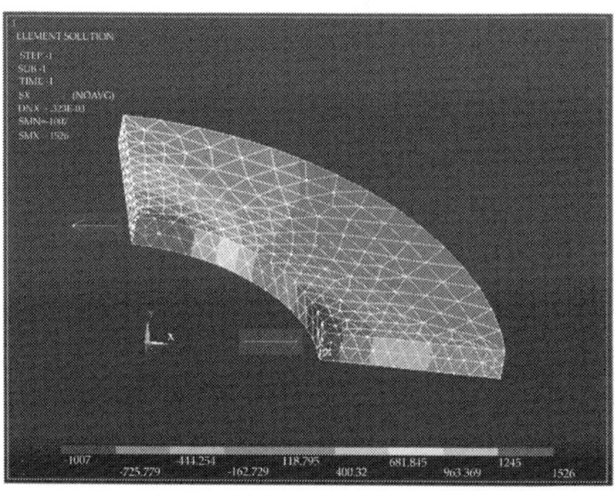

The theoretical solution for this problem is:

At the inside of the cylinder SX = -1000 psi and SZ = 1520.8 psi

At the outside of the cylinder SX = 0 psi and SZ = 520.8psi.

The upper arrow shows Sx normal to the Z-Y plane which corresponds to the cylinder hoop stress. The scale at the bottom indicates a maximum value of 1526 psi which is close to the theoretical value of 1521.

The lower arrow shows Sx in a direction normal to the inner surface of the cylinder with a scale value of -1007, close to the internal pressure of 1000 psi.

Thus the ANSYS calculated results seem to agree pretty well with the theory. But we need to examine the stresses more closely at the boundaries.

Number the nodes and elements.

10. PlotCntls -> Numbering . . .-> Turn on Node & Element Numbers Zoom in on the inside wall of the cylinder.

INSIDE ------------------------------------

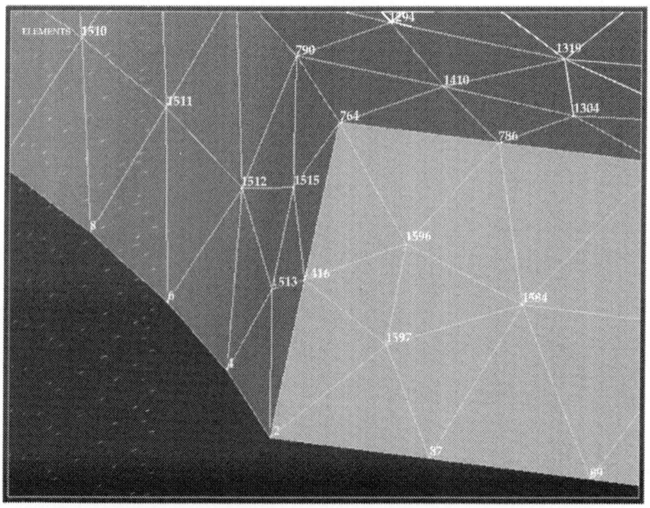

When the element stress solution contour plots are relatively smooth across element boundaries, as in this case, it indicates that the solution is reasonably accurate. Nodal Stress quantities can then be trusted as good representations of the solution.

List the stress results for node 764 .

11. List -> Results ->Nodal Solution . . . -> Stress -> Components -> OK.

THE FOLLOWING X,Y,Z VALUES ARE IN GLOBAL COORDINATES

NODE SX SY SZ SXY SYZ SXZ

764 -998.74 0.17087E-01 1521.7 0.98118E-01-0.50417 5.3997

Note that Sx = -998.74 is only 0.126 % different from the exact value of 1000, and Sz (hoop) = 1521.7 is only 0.06 % different from the theoretical value of 1520.8.

This problem can also be solved using brick elements. See the Mapped Meshes example for a guide to creating a mesh of 20 node brick elements for this cylindrical segment.

COMSOL

Introduction

COMSOL Multiphysics is afinite element analysis, solver and simulation software/FEA software package for various physics and engineering applications, especially coupled phenomena, or multiphysics. The package is cross-platform (Windows, Mac, Linux). In addition to conventional physics-based user interfaces, COMSOL Multiphysics also allows entering coupled systems of partial differential equations (PDEs). The PDEs can be entered directly or using the so-called weak form. Since version 5.0 (2014), COMSOL Multiphysics is also used for creating physics-based apps. These apps can be run with a regular COMSOL Multiphysics license but also with a COMSOL Server license. An early version (before 2005) of COMSOL Multiphysics was called FEMLAB.

Example

Mass Transfer

Consider a small drop of water, which is at the bottom of a vertical tunnel diffusing in the air contained within it. In order to accelerate the diffusion process, air is blown over the top of the

tunnel such that a negligible concentration of water at the top can be assumed. Strictly speaking this problem is transient; however, we can assume it as steady state to determine the concentration profile of water along the tunnel. Let us consider a spherical drop of water with a 1 mm radius as in Figure 1. To simplify the analysis, assume a perfectly cylindrical tunnel of 7.5 mm height and 2.5 mm diameter. The diffusion coefficient of water in air is 10-6 m2/s.

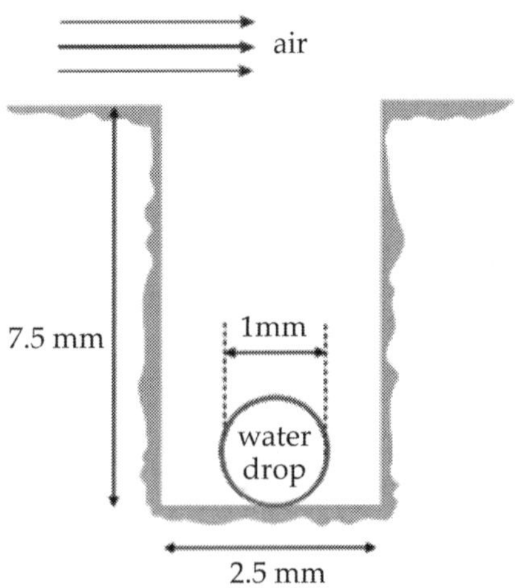

The region of interest is limited by the surface of the drop and the tunnel walls.

Start Up: Double click the Comsol 4.2 icon.

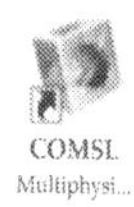

Model Wizard: The model wizard will appear in the interface; this will guide you through selecting the appropriate geometry, physics, and study type (stationary or time dependent). This problem can be described in just 2 dimensions because of the angular symmetry. Make sure you see why this problem is 2D despite dealing with 3D objects.

1) Select 2D and click the right facing arrow.

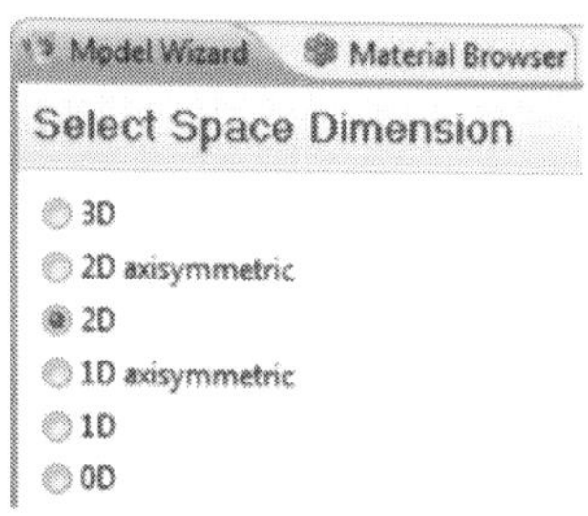

2) We now need to select the appropriate physics, in this case "transport of diluted species" will do the job. This can be found under the "Chemical species transport" tab.

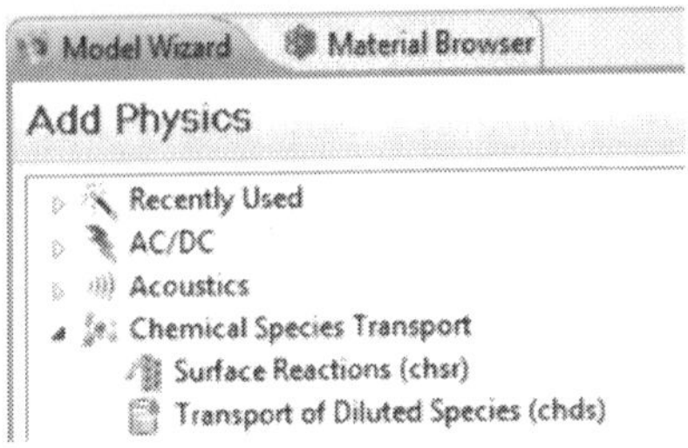

Finally we need to choose the study type. Because we are assuming steady state (even though this problem is inherently transient) the appropriate study type will be "stationary" To finish click the finish flag.

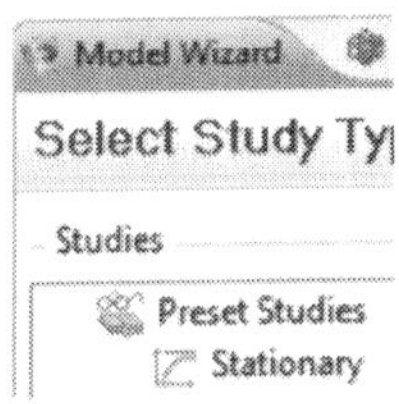

Geometry:

This geometry consists of both a rectangular region and a circular region. Accordingly we will need to add both a rectangle and a circle to create this geometry. Since the region we are really interested in is the region between the water droplet surface and the top of the cylindrical tunnel where water concentration is zero, we will actually subtract the circular water region from the rectangular air region using what is a called a Boolean operation. To do this:

1) Right click "geometry" and add a rectangle. Set height =7.5 [mm] and width=5 [mm]. The center of the rectangle should be at the origin.
2) Right click "geometry to add a circle. Set the radius = 1 [mm] and set the center position to x=0 y=-2.75 [mm]. After these two steps your geometry should look like the one below.

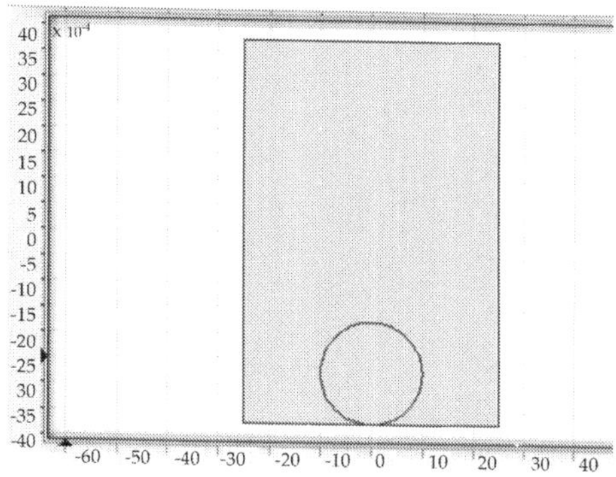

3) Now for the Boolean subtraction. Do this by right clicking on geometry, below where the different types input shapes are you will see "Boolean operations", highlight this to bring up the menu of different possible Boolean operations. In this case we want to select "difference.

4) Within the Boolean difference interface there are two boxes where different shapes can be entered. The first box is all the objects you wish to"add" while the second box contains the objects you wish to "subtract". We want to add the rectangle, but subtract the circle. To do this simply click the add box, then either left and right click the shape you wish to add to this box or left click to highlight then click the + button. Your interface should look like this once you finished.

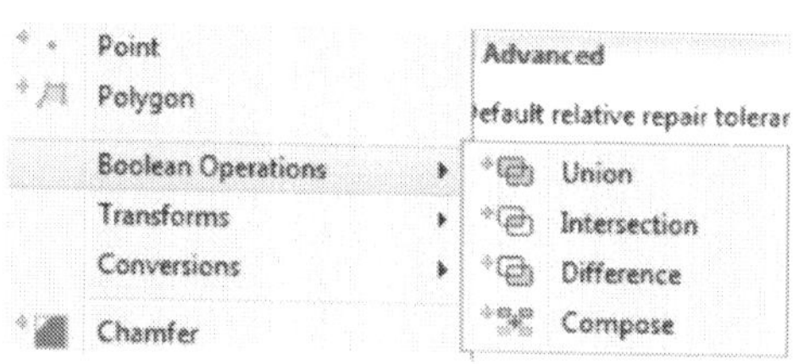

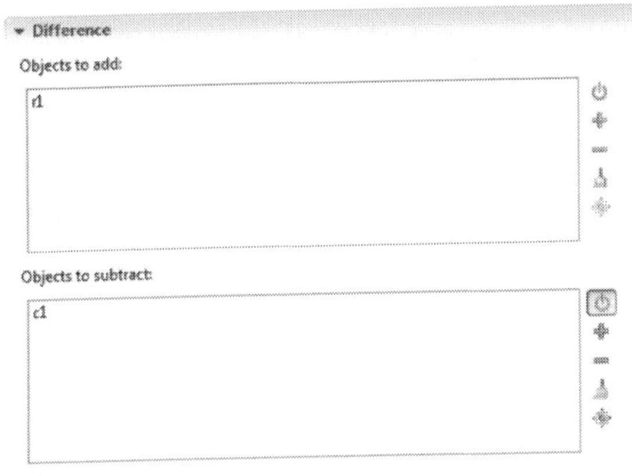

5) Once the proper shapes have been selected in the appropriate boxes click the "build all" button to carry out the subtraction. You should obtain the following result

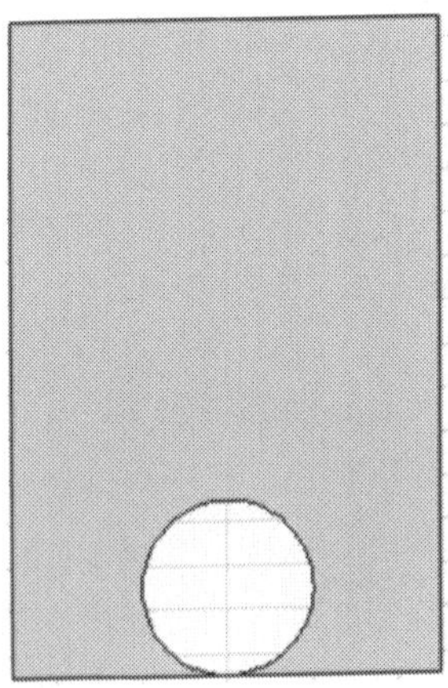

Transport of Diluted Species:

Here we need only do 2 things. Set the diffusivity of water hrough air and set the boundary conditions.

- Under the Transport of Diluted Species tab you will see the "convection and diffusion" tab.

  Left click this to open the interface.

  You will see a "Diffusion" tab within this interface

here you can set Dc equal to the appropriate value of $1\times10^{-6}m^2/s$. This should look like the graphic below.

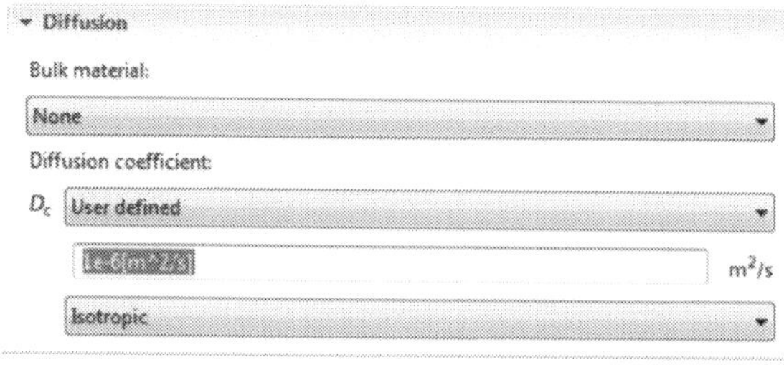

- We now need to input the appropriate boundary conditions to the problem. Since our two conditions are values of concentration at two surfaces we will need to add two "concentration"boundary conditions. The concentration boundary condition is found by right clicking the Transport of Diluted Species tab and left clicking "concentration", since we have two boundaries we will need two separate boundary conditions.

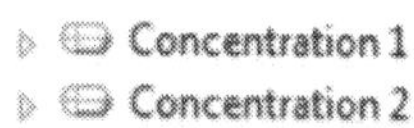

- First let us add the boundary at the surface of the water droplet. This surface will have the concentration of pure water thus set $c = 55556 \text{ mol}/m^3$. This condition must be applied to the appropriate subdomain so in the usual manner, left click to highlight and right click to add the appropriate surface while in the concentration 1 interface.

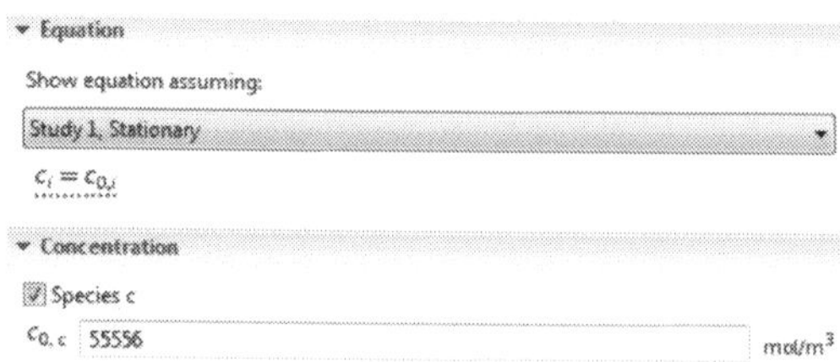

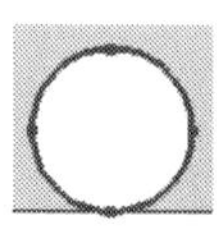

- Now in similar fashion we need to add the negligible concentration of water condition to the top surface. At this surface we can assume c=0.

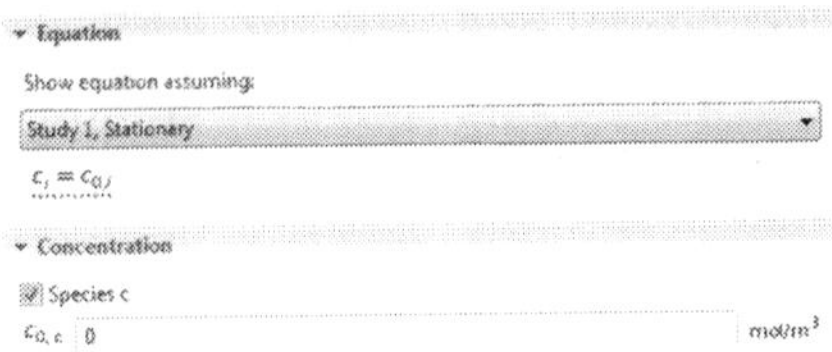

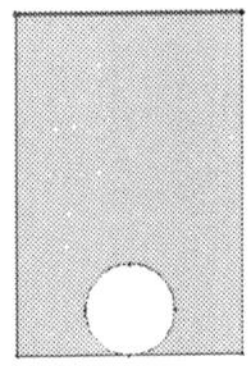

For a simple diffusion problem such as this specifying the appropriate boundary conditions and the appropriate diffusivity (the only physical parameter which comes directly into play in diffusion) is all we need to do. The problem is now ready to be solved as the meshing by default will be fine for a problem like this. Right click on study and left click compute in the menu it brings up.

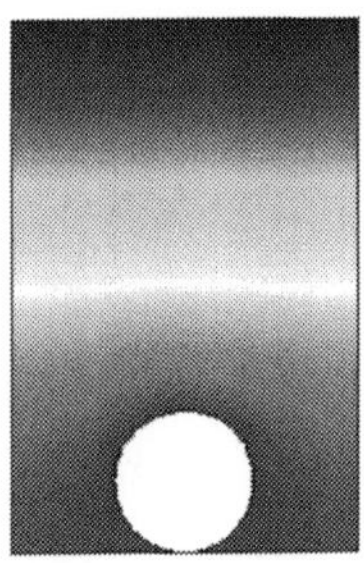

Results:

Now that we have our solution let us examine it using two important techniques with which you

should be familiar.

2D Cut Line:

- First we will examine the concentration profile of water 1 mm away from the wall. This is done by right clicking on "data sets" under the results tab and left clicking on "Cut Line 2D".

  This will add a new data set under the data set tab, however the two points defining this line segment need to be specified. The points x1=-1.5 [mm] y1=-3.75 [mm] and x2=-1.5 [mm]

y2=3.75 [mm].

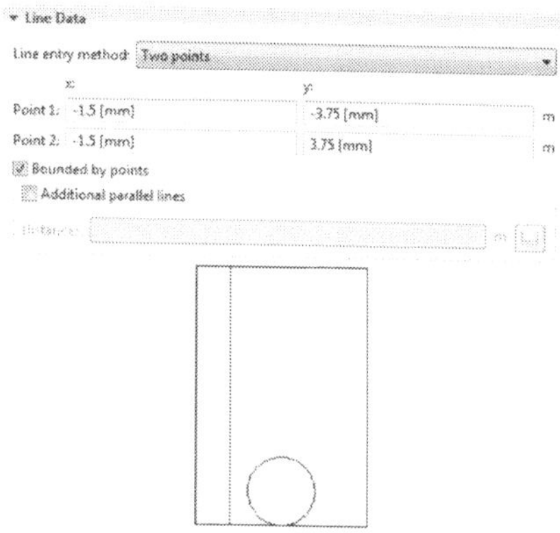

- We now need to add a 1D plot group to display this cut line data. Right click on Results and left click "1D Plot Group", this will add the one dimensional plot group where the data can be displayed. Right click the "1D plot group" tab and select "Line Graph", finally in the "Line Graph" interface choose "Cut line 2D" in the data set box and click the paint brush (graph button at top right of interface).

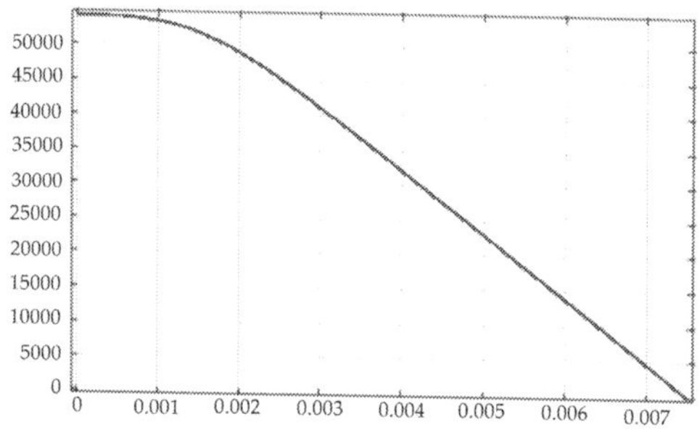

The second graphical tool we will employ is called a contour plot. This type of plot connects all points of an equal value with a certain parameter difference between each line. The closer together the lines the faster the change in whatever parameter is being contoured.

Contour Plot of Concentration:

- To graph the surface contours of concentration we will need to add another 2D plot group. Do this the same way as adding a 1D plot group, right click on results and left click 2D plot group.

- Now that the second 2D plot group is up right click this tab and select "Contour". This will bring up the contour plot interface, by default the parameter to be contoured is concentration c(concentration of water). All that remains to do is to press the plot button (the paint brush button again at the upper right of the interface).

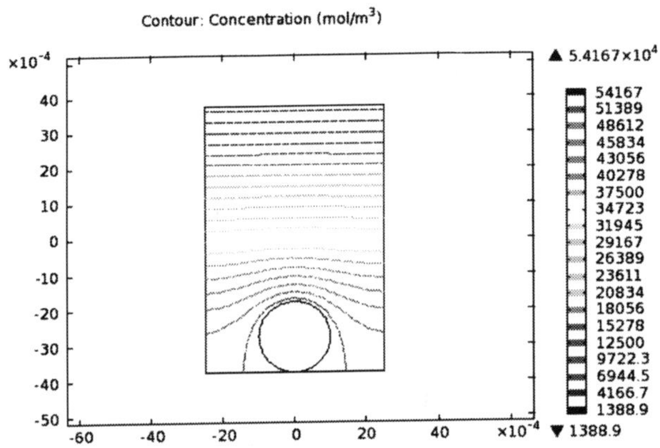

The application of COMSOL also have another example here to show the heat problem. This example is a model od a continuous casting process.

Description

Liquid metal is poured into a mold of uniform cross section. The outside of the mold is cooled and the metal solidifies as it flows through the mold. When the metal leaves the mold, it is completely solidified on the outside but still liquid inside. The metal then continues to cool and eventually solidify completely, at which point it can be cut into sections.

This model will not calculate the flow field, and there is two assumptions that there is no volume change during solidification and the velocity of the metal is constant.

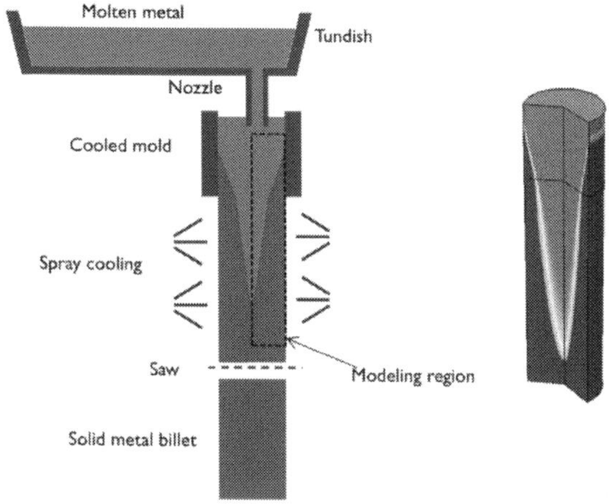

Figure 4.12 Illustration of the casting process

This model simplifies the 3D geometry of the casting to a 2D axisymmetric model which has two regions: one representing the strand within the mold, the other one spray cooled region outside of the mold.

Since this a continuous process, the system can be modeled at steady state. The heat transport is described by the equation

$$\nabla \cdot (-k\nabla T) = -\ C_p u \nabla T$$

Here, k and C_p denote thermal conductivity and specific heat, respectively. The velocity ,u is fixed casting speed of the metal in both liquid and solid state.

As the metal cools down in the mole, it solidifies. During the phase transition, as significant amount of latent heat is released. The total amount of the heart released per unit mass of alloy during the transition is given by the change in enthalpy, ΔH. In addition, the specific heat capacity, C_p, also changes considerably during the transition. The difference in specific heat before and after transition can be approximated by

$$\Delta C_p = \frac{\Delta H}{T}$$

As opposed to pure metals, an alloy generally undergoes a broad temperature transition zone, over several Kelvin, in which a mixture of both solid and molten material co-exist in a “mushy” zone. To account for the latent heat related to the phase transition, the Apparent Heat capacity method is used through the Heat Transfer with Phase Change domain condition. The objective of the analysis is to make ΔT, the half-width of the transition interval small, such that the solidification front location is well defined.

This example is a highly nonlinear problem and benefits from taking an iterative approach to finding the solution. The location of the transition between the molten and solid state is a strong function of the casting velocity, the cooling rate in the mold, and the collign rate in the spray cooled region. A fine mesh is needed across the solidification front to resolve the change in material properties. However, it is not known where this front will be.

By starting with a gradual transition between liquid and solid, it is possible to find a solution even on a relatively coarse mesh. This solution can be used as the starting point for the next step in the solution procedure, which uses a sharper transition from liquid to solid. This is done using the continuation method. Given a monotonic list of values to solve for, the continuation method uses the solution to the last case as the starting condition for the next. Once a solution is found for the smallest desired ΔT, the adaptive mesh refinement algorithm is used to refine the mesh to put more elements around the transition region. This finer mesh is then used to find a solution with an even sharper transition. This can be repeated as needed to get better and better resolution of the locatin of the solidification front.

In this example, the parameter ΔT is first ramped down from 300K to 75K, then the adaptive mesh refinement is used such that a finer mesh is used around the solidification front. The resultant solution and mesh are then used as starting points for second study, where the paramenter ΔT is further ramped down from 50K to 25 K. The double-dogleg solver is used to find the solution to this highly nonlinear problem. Although it takes more time, this solver converges better in cases when material properties vary strongly with respect to the solution.

The solidification front computed with the coarsest mesh, and for

$\Delta T = 75K$, is shown in Figure 4.13. A wide transition between the molten and solid state is observed. The adaptive mesh refinement algorithm then refines the mesh along the solidification front because this is the region where the results are strongly dependent upon mesh size. This solution, and refined mesh is used as the starting point for the next solution, which ramps the ΔT parameter down to 25 K. These results are shown in Figure 4.14.

The point of complete solidification moves slightly as the transition zone is made smaller. As the transition zone becomes smaller, a finer mesh is needed, otherwise the model might not converge. If it is desired to get an even better resolution of the solidification front, the solution procedure used here should be repeated to get an even finer mesh, and further ramp down the ΔT parameter.

The liquid phase fraction is plotted along the r-direction at the line at the bottom of the mold in Figure 4.13 . Figure 4.14 shows the liquid fraction along the centerline of the strand. For smaller values of ΔT, the transition becomes sharper, and the model gives confidence that the metal is completely solidified before the strand is cut.

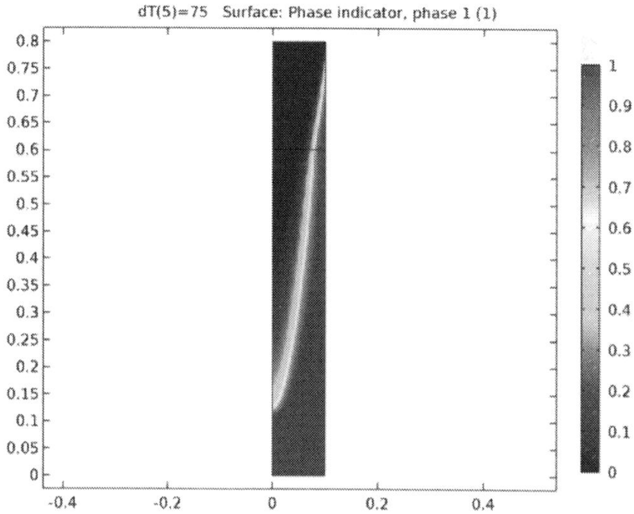

Figure 4.13 The fraction of liquid phase for $\Delta T = 75K$ shows a gradual transition between the liquid and solid phase.

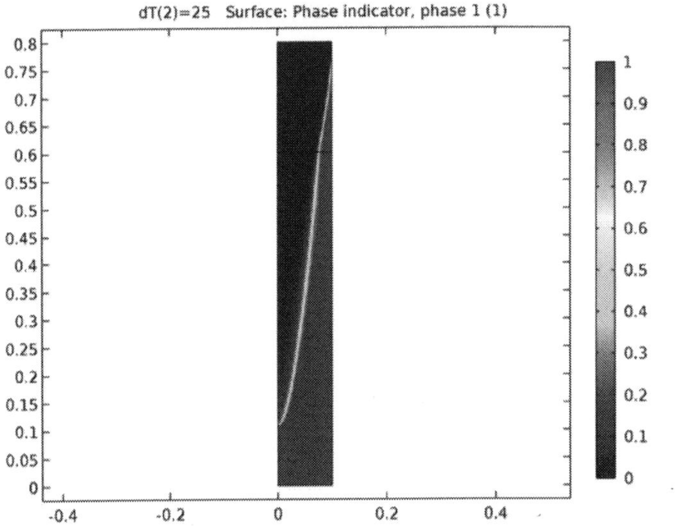

Figure 4.14 The fraction of liquid phase for $\Delta T = 25K$ shows a sharp transition between the liquid and solid phase

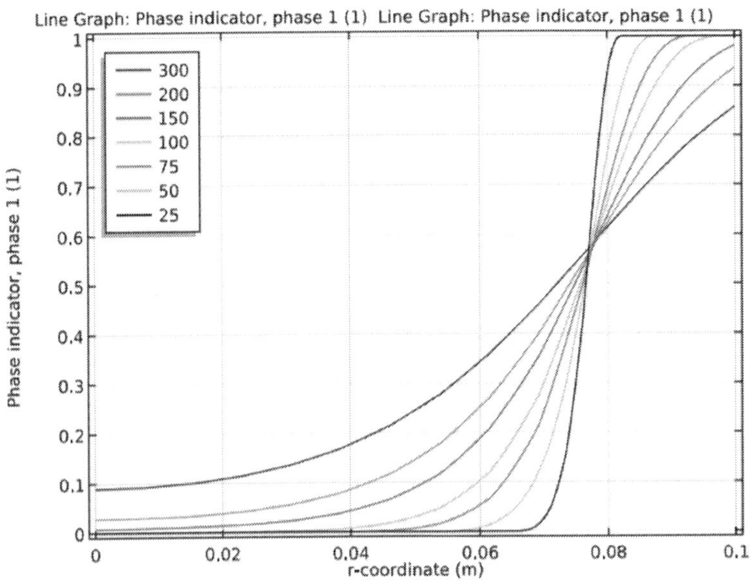

Figure 4.15 The fraction of liquid phase through the radius for all values of ΔT. For smaller value of ΔT, the transition is sharper

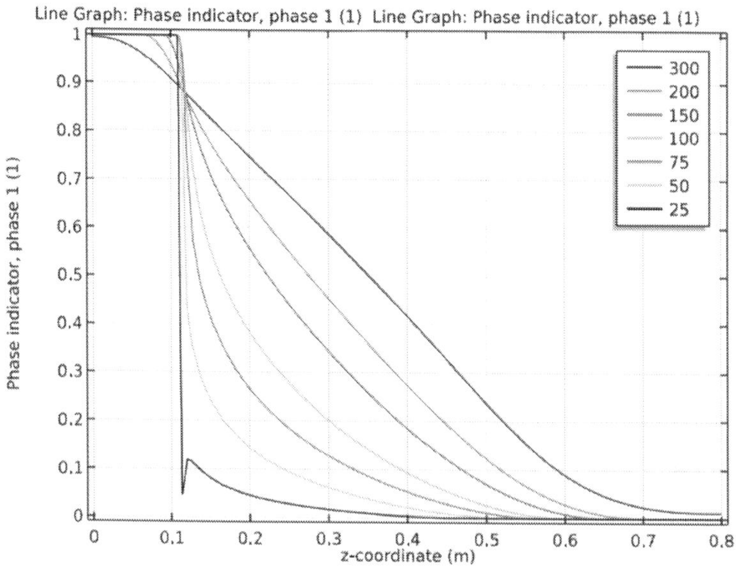

Figure 4.16 The fraction of liquid phase along the centerline for all values of ΔT . For smaller value of ΔT , the transition is sharper.

Detail

Modeling Instructions

New

1 Click Model Wizard

Model Wizard

1 In the Model Wizard window, click 2D axisymmetric

2 In the select physics tree, select Heat Transfer, Heat transfer in Fluids

3 Click Add

4 Click Study

5 In the select study tree, select Present Studies, stationary

6 Click Done

Global Definitions

First, set up parameters and variables used in the continuous casting process model analysis.

Parameters

1 On the Home toolbar, click Parameters.

2 In the Settings window for Parameters, locate the Parameters section.

3 Click Load from File.

4 Browse to the application's Application Library folder and double-click the file "cooling_solidification_metal_parameters.txt".

Geometry

Create two rectangles representing the strand within the mold, and spray cooled region outside of the mold.

Rectangle 1 (r1)

1 On the Geometry toolbar, click Primitives and choose Rectangle.

2 In the Settings window for Rectangle, locate the Size section.

3 In the Width text field, type 0.1.

4 In the Height text field, type 0.6.

Rectangle 2 (r2)

1 On the Geometry toolbar, click Primitives and choose Rectangle.

2 In the Settings window for Rectangle, locate the Size section.

3 In the Width text field, type 0.1.

4 In the Height text field, type 0.2.

5 Locate the Position section. In the z text field, type 0.6.

6 Click the Build All Objects button

7 Click the Zoom Extents button on the Graphics toolbar.

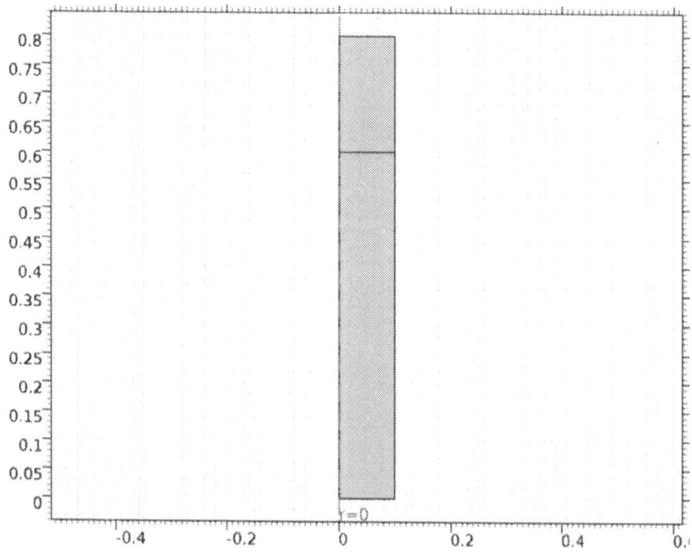

Materials

Material 1 (mat1)

1 In the Model Builder window, under Component 1 (comp1) right-click Materials and choose Blank Material.

2 In the Settings window for Material, type Solid Metal Alloy in the Label text field.

3 Locate the Material Contents section. In the table, enter the following settings:

Property	Name	Value	Unit	Property group
Thermal conductivity	k	300	W/(m·K)	Basic
Density	rho	8500	kg/m³	Basic
Heat capacity at constant pressure	Cp	Cp_s	J/(kg·K)	Basic
Ratio of specific heats	gamma	1	1	Basic

Material 2 (mat2)

1 In the Model Builder window, right-click Materials and choose Blank Material.

2 In the Settings window for Material, type Liquid Metal Alloy in the

Label text field.

3 Select Domains 1 and 2 only

4 Locate the Material Contents section. In the table, enter the following settings:

Property	Name	Value	Unit	Property group
Thermal conductivity	k	150	W/(m·K)	Basic
Density	rho	8500	kg/m³	Basic
Heat capacity at constant pressure	Cp	Cp_1	J/(kg·K)	Basic
Ratio of specific heats	gamma	1	1	Basic

Heat Transfer

Initial Values 1

1 In the Model Builder window, under Component 1 (comp1)>Heat Transfer in Fluids (ht), click Initial Values 1.

2 In the Settings window for Initial Values, locate the Initial Values section.

3 In the T text field, type T_in.

Heat Transfer with Phase Change 1

1 On the Physics toolbar, click Domains and choose Heat Transfer with Phase Change.

2 Select Domains 1 and 2 only.

3 In the Settings window for Heat Transfer with Phase Change, locate the Model Inputs section.

4 Specify the u vector as

0	r
-v_cast	z

5 Locate the Phase Change section. In the Tpc,1 → 2 text field, type T_m.

6 In the ΔT1 → 2 text field, type dT.

7 In the L1 → 2 text field, type dH.

8 Locate the Phase 1 section. From the Material, phase 1 list, choose Solid Metal Alloy

(mat1).

9 Locate the Phase 2 section. From the Material, phase 2 list, choose Liquid Metal Alloy

(mat2).

Temperature 1

1 On the Physics toolbar, click Boundaries and choose Temperature.

2 Select Boundary 5 only.

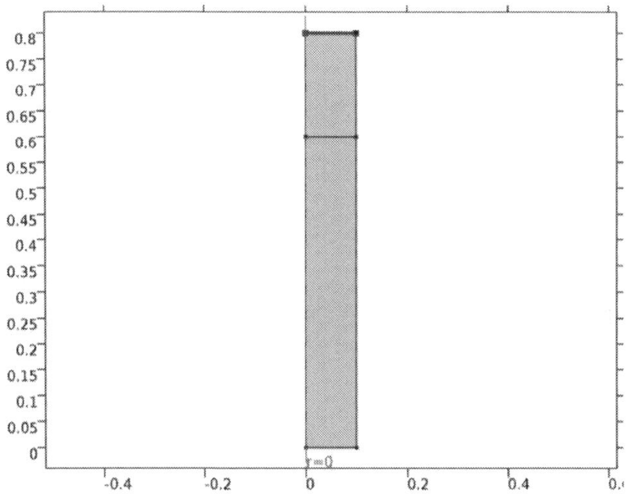

3 In the Settings window for Temperature, locate the Temperature section.

4 In the T0 text field, type T_in.

Heat Flux 1

1 On the Physics toolbar, click Boundaries and choose Heat Flux.

2 Select Boundary 7 only.

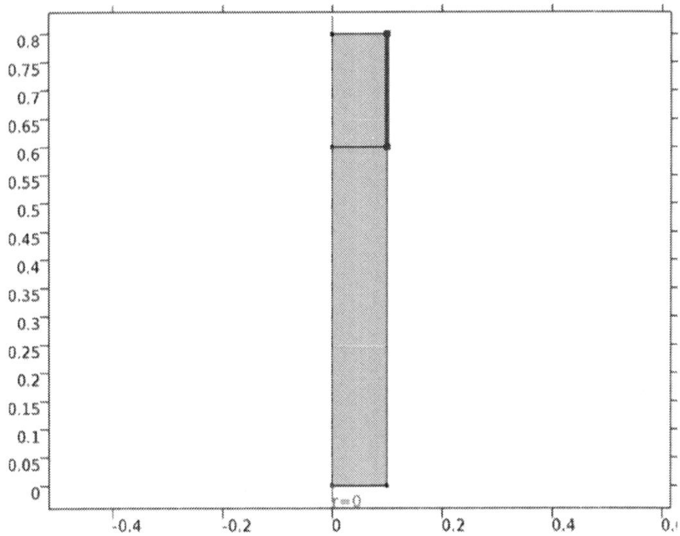

3 In the Settings window for Heat Flux, locate the Heat Flux section.

4 Click the Convective heat flux button.

5 In the h text field, type h_mold.

6 In the Text text field, type T0.

Heat Flux 2

1 On the Physics toolbar, click Boundaries and choose Heat Flux.

2 Select Boundary 6 only.

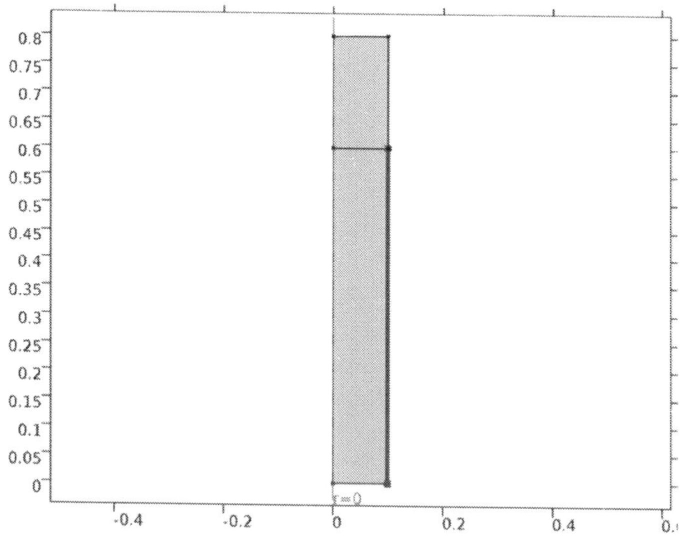

3 In the Settings window for Heat Flux, locate the Heat Flux section.

4 Click the Convective heat flux button.

5 In the h text field, type h_spray.

6 In the Text text field, type T0.

Diffuse Surface 1

1 On the Physics toolbar, click Boundaries and choose Diffuse Surface.

2 Select Boundary 6 only.

3 In the Settings window for Diffuse Surface, locate the Surface Emissivity section.

4 From the ε list, choose User defined. In the associated text field, type eps_s.

5 Locate the Ambient section. In the Tamb text field, type T0.

Outflow 1

1 On the Physics toolbar, click Boundaries and choose Outflow.

2 Select Boundary 2 only

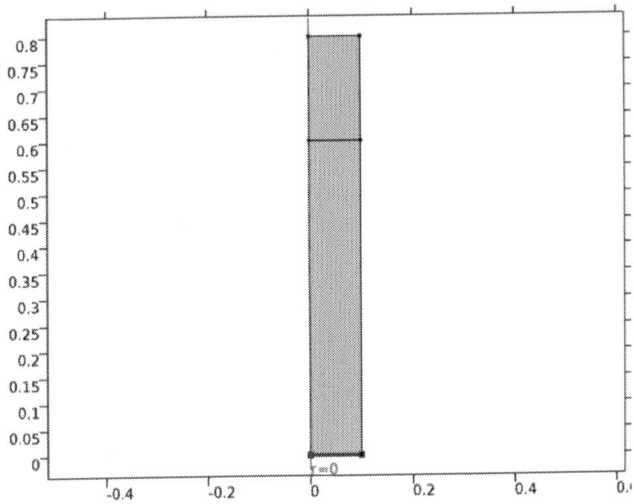

MESH 1

1 In the Model Builder window, under Component 1 (comp1) click Mesh 1.

2 In the Settings window for Mesh, locate the Mesh Settings section.

3 From the Element size list, choose Extremely fine.

STUDY 1

1 In the Model Builder window, click Study 1.

2 In the Settings window for Study, locate the Study Settings section.

3 Clear the Generate default plots check box.

Step 1: Stationary

Set up an auxiliary continuation sweep for the dT parameter.

1 In the Model Builder window, under Study 1 click Step 1: Stationary.

2 In the Settings window for Stationary, click to expand the Study extensions section.

3 Locate the Study Extensions section. Select the Auxiliary sweep check box.

4 Click Add.

5 In the table, enter the following settings:

Parameter name	Parameter value list	Parameter unit
dT	300 200 150 100 75	

6 Select the Adaptive mesh refinement check box

Solution 1

1 On the Study toolbar, click Show Default Solver.

2 In the Model Builder window, expand the Solution 1 node.

3 In the Model Builder window, expand the Study 1>Solver Configurations>Solution

1>Stationary Solver 1 node, then click Adaptive Mesh Refinement.

4 In the Settings window for Adaptive Mesh Refinement, locate the Mesh Refinement section.

5 From the Refinement method list, choose Mesh initialization.

6 In the Model Builder window, under Study 1>Solver Configurations>Solution1>Stationary Solver 1 click Fully Coupled 1.

7 In the Settings window for Fully Coupled, click to expand the Method and termination section.

8 Locate the Method and Termination section. From the Nonlinear method list, choose Double dogleg.

9 On the Study toolbar, click Compute.

RESULTS

2D Plot Group 1

1 On the Home toolbar, click Add Plot Group and choose 2D Plot Group.

2 In the Settings window for 2D Plot Group, type Solid and Liquid Phases

(Adaptive Mesh) in the Label text field.

Solid and Liquid Phases (Adaptive Mesh)

1 Right-click Results>Solid and Liquid Phases (Adaptive Mesh) and choose Surface.

2 In the Settings window for Surface, click Replace Expression in the upper-right corner

of the Expression section. From the menu, choose Model>Component 1>Heat Transfer

in Fluids>Phase change>ht.theta1 - Phase indicator, phase 1.

3 On the Solid and Liquid Phases (Adaptive Mesh) toolbar, click Plot.

The reproduced figure describes the fraction of liquid phase for ÄT = 75 K .

ADD STUDY

1 On the Home toolbar, click Add Study to open the Add Study window.

2 Go to the Add Study window.

3 Find the Studies subsection. In the Select study tree, select Preset Studies.

4 In the Select study tree, select Preset Studies>Stationary.

5 Click Add Study in the window toolbar

STUDY 2

Step 1: Stationary

1 On the Home toolbar, click Add Study to close the Add Study window.

2 In the Model Builder window, click Study 2.

3 In the Settings window for Study, locate the Study Settings section.

4 Clear the Generate default plots check box.

5 In the Model Builder window, click Step 1: Stationary.

6 In the Settings window for Stationary, click to expand the Values of dependent

variables section.

7 Locate the Values of Dependent Variables section. Select the Initial values of variables solved for check box.

8 From the Method list, choose Solution.

9 From the Study list, choose Study 1, Stationary.

10 From the Solution list, choose Adaptive Mesh Refinement 1.

11 From the Parameter value (dT) list, choose 75.

12 Click to expand the Mesh selection section. Locate the Study Extensions section. Select the Auxiliary sweep check box.

13 Click Add.

14 In the table, enter the following settings:

Parameter name	Parameter value list	Parameter unit
dT	50 25	

Solution 3

1 On the Study toolbar, click Show Default Solver.

2 In the Model Builder window, expand the Solution 3 node.

3 In the Model Builder window, expand the Study 2>Solver Configurations>Solution 3>Stationary Solver 1 node, then click Fully Coupled 1.

4 In the Settings window for Fully Coupled, locate the Method and Termination section.

5 From the Nonlinear method list, choose Double dogleg.

6 On the Study toolbar, click Compute.

RESULTS

2D Plot Group 2

1 On the Home toolbar, click Add Plot Group and choose 2D Plot Group.

2 In the Settings window for 2D Plot Group, type Solid and Liquid Phases in the

Label text field.

3 Locate the Data section. From the Data set list, choose Study 2/Solution 3.

Solid and Liquid Phases

1 Right-click Results>Solid and Liquid Phases and choose Surface.

2 In the Settings window for Surface, click Replace Expression in the upper-right corner of the Expression section. From the menu, choose Model>Component 1>Heat Transfer in Fluids>Phase change>ht.theta1 - Phase indicator, phase 1.

3 On the Solid and Liquid Phases toolbar, click Plot. This shows the fraction of liquid phase for $\Delta T = 25$ K.

1D Plot Group 3

1 On the Home toolbar, click Add Plot Group and choose 1D Plot Group.

2 In the Settings window for 1D Plot Group, type Phase Indicator at Symmetry

Axis in the Label text field.

Phase Indicator at Symmetry Axis

1 On the Phase Indicator at Symmetry Axis toolbar, click Line Graph.

2 Select Boundaries 1 and 3 only.

3 In the Settings window for Line Graph, locate the y-axis data section.

4 Click ht.theta1 - Phase indicator, phase 1 in the upper-right corner of the section.

Locate the x-Axis Data section. From the Parameter list, choose Expression.

5 In the Expression text field, type z.

6 Click to expand the Legends section. Select the Show legends check box.

7 Right-click Results>Phase Indicator at Symmetry Axis>Line Graph 1 and choose

Duplicate.

8 In the Settings window for Line Graph, locate the Data section.

9 From the Data set list, choose Study 2/Solution 3.

10 On the Phase Indicator at Symmetry Axis toolbar, click Plot.

Compare the resulting plot with Figure 5 showing the fraction of liquid phase

through the centerline for all values of ΔT.

Phase Indicator at Symmetry Axis 1

1 In the Model Builder window, right-click Phase Indicator at Symmetry Axis and choose

Duplicate.

2 In the Settings window for 1D Plot Group, type Phase Indicator through

Radius in the Label text field.

Phase Indicator through Radius

1 In the Model Builder window, expand the Results>Phase Indicator through Radius node, then click Line Graph 1.

2 In the Settings window for Line Graph, locate the Selection section.

3 Click Clear Selection.

4 Select Boundary 4 only.

5 Locate the x-Axis Data section. In the Expression text field, type r.

6 On the Phase Indicator through Radius toolbar, click Plot.

7 In the Model Builder window, under Results>Phase Indicator through Radius click Line

Graph 2.

8 In the Settings window for Line Graph, locate the Selection section.

9 Click Clear Selection.

10 Select Boundary 4 only.

11 Locate the x-Axis Data section. In the Expression text field, type r.

12 In the Model Builder window, click Phase Indicator through Radius.

13 In the Settings window for 1D Plot Group, click to expand the Legend section.

14 From the Position list, choose Upper left.

15 On the Phase Indicator through Radius toolbar, click Plot.

Compare the resulting plot with Figure 4.14 showing the fraction of liquid phase through the radius for all values of ÄT .

CHAPTER 4

FINITE ELEMENT METHOD IN SOLID MECHANICS

LINEAR PROBLEM

Ideal Material Response

Conducting the stretch test on one sample with pretty small force, the following behavior will be observed:

(1) if the loads on the sample are missing, this sample no matter what materials will spring back to the old state, in another way, instead of axial direction, it will shrink the opposite direction;

(2) As is known to all, stress usually increase or decrease by strain. The relationship between these two is linear.

(3) Stress determines the variation of strain, even the pressure speed and the before pressure either cause the change of strain;

(4) The definition of "isotropic", the stretch direction of the sample in axial tensile test may be not uniform, but the

relationship between stress and strain is always same with each other. This related with the material property of the sample, for the material which has such peoperty, called isotropic material;

(5) If the material is isotropic (no preferred material orientation) and homogeneous, then the specimen will simply increase in size, without shape change.

(6) If the material is heated to certain degree, the temperature of the sample will also gained regularly, the original shape of the sample will gently effected by the heat.

Index of Ideal Elastic Material in Mechanical and Thermal Property.

The term of strain and stress can presented out in varies forms. Here only consider small shape changes situation, so the first step is to determine these two important terms.

Infinitesimal strain tensor is shown below:

$$\varepsilon_{ij} = \frac{\left(\dfrac{\partial u_i}{\partial x_j} + \dfrac{\partial u_j}{\partial x_i}\right)}{2}$$

Small shape change and rotation are the only conditions to use this equation. This form is good for calculating results.

So the Cauchy stress σ_{ij} is used to match the Cauchy strain ε_{ij} as well.

The next step is to thinking about stress-strain relationship. The relationship is shown by the follow equations.

Note: All the solid material considered here are linear, elastic, isotropic solid materials.

$$\begin{bmatrix}\varepsilon_{11}\\ \varepsilon_{22}\\ \varepsilon_{33}\\ 2\varepsilon_{23}\\ 2\varepsilon_{13}\\ 2\varepsilon_{12}\end{bmatrix}=\frac{1}{E}\begin{bmatrix}1 & -\upsilon & -\upsilon & 0\ 0\ 0 & & \\ -\upsilon & 1 & -\upsilon & 0\ 0\ 0 & & \\ -\upsilon & -\upsilon & 1 & 0\ 0\ 0 & & \\ 0\ 0\ 0 & & & 2(1+\upsilon) & 0 & 0\\ 0\ 0\ 0 & & & 0 & 1(1+\upsilon) & 0\\ 0\ 0\ 0 & & & 0 & 0 & 2(1+\upsilon)\end{bmatrix}\begin{bmatrix}\sigma_{11}\\ \sigma_{22}\\ \sigma_{33}\\ \sigma_{23}\\ \sigma_{13}\\ \sigma_{12}\end{bmatrix}+\alpha\Delta T\begin{bmatrix}1\\1\\1\\0\\0\\0\end{bmatrix}$$

E- Young's modulus

v- Poisson's ratio

α - coefficient of thermal expansion

ΔT-increase in temperature of the solid

σ_{ij} and ε_{ij} are symmetric, it can be deduced from the above equation.

Another stress-strain relationship can be written as:

$$\begin{bmatrix}\sigma_{11}\\ \sigma_{22}\\ \sigma_{33}\\ \sigma_{23}\\ \sigma_{13}\\ \sigma_{12}\end{bmatrix}=\frac{E}{(1+\nu)(1-2\upsilon)}\begin{bmatrix}1-\upsilon & \upsilon & \upsilon & & 0\ 0\ 0 & \\ \upsilon & 1-\upsilon & \upsilon & & 0\ 0\ 0 & \\ \upsilon & \upsilon & 1-\upsilon & & 0\ 0\ 0 & \\ & & & \frac{(1-2\upsilon)}{2} & 0 & 0\\ & 0\ 0\ 0 & & 0 & \frac{(1-2\upsilon)}{2} & 0\\ & 0\ 0\ 0 & & & & \\ & 0\ 0\ 0 & & 0 & 0 & \frac{(1-2\upsilon)}{2}\end{bmatrix}\begin{bmatrix}\varepsilon_{11}\\ \varepsilon_{22}\\ \varepsilon_{33}\\ 2\varepsilon_{23}\\ 2\varepsilon_{13}\\ 2\varepsilon_{12}\end{bmatrix}$$

$$+\frac{E\alpha\Delta T}{1-2v}\begin{bmatrix}1\\1\\1\\0\\0\\0\end{bmatrix}$$

Notes: The coefficient 2 of strain vector is not always same. It is only true for isotropic materials. If the sample is anisotropic, this coefficient is not necessary be 2. Although many text books and code use 2 as a constant. It should be aware of.

The above expression is equivalent with the bellow one , this form using index notation is more convenient.

$$\varepsilon_{ij} = \frac{1+\nu}{E}\sigma_{ij} - \frac{\nu}{E}\sigma_{kk}\delta_{ij} + \alpha\Delta T\delta_{jj}$$

Another expression is

$$\sigma_{ij} = \frac{E}{1+\nu}\left\{\varepsilon_{ij} + \frac{\nu}{1-2\nu}\varepsilon_{kk}\delta_{ij}\right\} + \frac{E\alpha\Delta T}{1-2\nu}\delta_{jj}$$

Another expression for stress strain relationship is shown bellow by C_{ijkl} and S_{ijkl}, which are elastic modulus tensor and elastic compliance tensor separately.

$$\sigma_{ij} = C_{ijkl}\left(\varepsilon_{kl} - \alpha\Delta T\delta_{kl}\right)$$

$$\varepsilon_{ij} = S_{ijkl}\sigma_{kl} + \alpha\Delta T\delta_{kl}$$

The definition of C_{ijkl} and S_{ijkl} are:

$$C_{ijkl} = \frac{E}{2(1+\upsilon)}\left\{\delta_{ij}\delta_{jk} + \delta_{ik}\delta_{jl}\right\} + \frac{E\upsilon}{(1+\upsilon)(1-2\nu)}\delta_{ij}\delta_{kl}$$

$$S_{ijkl} = \frac{1+\nu}{2E}\left(\delta_{ij}\delta_{jk} + \delta_{ik}\delta_{jl}\right) - \frac{\nu}{E}\delta_{ij}\delta_{kl}$$

Planar deformation's reduced constitutive equations. (Ideal material)

Stress-strain relationship can also be simplified by special situation, for example, if the deformation is planar, the components of strain and stress are zeros in the matrix form. Thus it is easier to calculate.

For example, $\varepsilon_{33} = \varepsilon_{23} = \varepsilon_{13} = 0$. ó − å relationship are turn to be:

$$\begin{bmatrix} \varepsilon_{11} \\ \varepsilon_{22} \\ 2\varepsilon_{12} \end{bmatrix} = \frac{(1+\nu)}{E} \begin{bmatrix} 1-\nu & -\nu & 0 \\ -\nu & 1-\nu & 0 \\ 0 & 0 & 2 \end{bmatrix} \begin{bmatrix} \sigma_{11} \\ \sigma_{22} \\ \sigma_{12} \end{bmatrix} + (1+\nu)\alpha\Delta T \begin{bmatrix} 1 \\ 1 \\ 0 \end{bmatrix}$$

$$\begin{bmatrix} \sigma_{11} \\ \sigma_{22} \\ \sigma_{12} \end{bmatrix} = \frac{E}{(1+\nu)(1-2\nu)} \begin{bmatrix} 1-\nu & \nu & 0 \\ \nu & 1-\nu & 0 \\ 0 & 0 & \dfrac{1-2\nu}{2} \end{bmatrix} \begin{bmatrix} \varepsilon_{11} \\ \varepsilon_{22} \\ 2\varepsilon_{12} \end{bmatrix} + \frac{E\alpha\Delta T}{1-2\nu} \begin{bmatrix} 1 \\ 1 \\ 0 \end{bmatrix}$$

$$\sigma_{33} = \frac{E\nu(\varepsilon_{11}+\varepsilon_{22})}{(1+\nu)(1-2\nu)} + \frac{E\alpha\Delta T}{1-2\nu}$$

$$\sigma_{13} = \sigma_{23} = 0$$

Another expression:

$$\varepsilon_{\alpha\beta} = \frac{1+\nu}{E}\left(\sigma_{\alpha\beta} - \nu\sigma_{rr}\delta_{\alpha\beta}\right) + (1+\nu)\alpha\Delta T\delta_{\alpha\beta}$$

$$\sigma_{\alpha\beta} = \frac{E}{1+\nu}\left(\varepsilon_{\alpha\beta} + \frac{\nu}{1-2\nu}\varepsilon_{rr}\delta_{\alpha\beta}\right) - \frac{E\alpha\Delta T}{1-2\nu}\delta_{\alpha\beta}$$

Here á,â -1 or 2.

Planar stress deformation has such expression.

$$\sigma_{33} = \sigma_{13} = \sigma_{23} = 0$$

$$\begin{bmatrix} \varepsilon_{11} \\ \varepsilon_{22} \\ 2\varepsilon_{12} \end{bmatrix} = \frac{1}{E} \begin{bmatrix} 1 & -\nu & 0 \\ -\nu & 1 & 0 \\ 0 & 0 & 2(1+\nu) \end{bmatrix} \begin{bmatrix} \sigma_{11} \\ \sigma_{22} \\ \sigma_{12} \end{bmatrix} + \alpha\Delta T \begin{bmatrix} 1 \\ 1 \\ 0 \end{bmatrix}$$

$$\begin{bmatrix} \sigma_{11} \\ \sigma_{22} \\ \sigma_{12} \end{bmatrix} = \frac{E}{\left(1-\nu^2\right)} \begin{bmatrix} 1 & \nu & 0 \\ \nu & 1 & 0 \\ 0 & 0 & \frac{1-\nu}{2} \end{bmatrix} \begin{bmatrix} \varepsilon_{11} \\ \varepsilon_{22} \\ 2\varepsilon_{12} \end{bmatrix} + \frac{E\alpha\Delta T}{1-\nu} \begin{bmatrix} 1 \\ 1 \\ 0 \end{bmatrix}$$

$$\varepsilon_{33} = \frac{-\nu\left(\sigma_{11}+\sigma_{22}\right)}{E} + \alpha\Delta T$$

$$\varepsilon_{\alpha\beta} = \frac{1+\nu}{E}\left(\sigma_{\alpha\beta} - \frac{\nu}{1+\nu}\sigma_{rr}\delta_{\alpha\beta}\right) + \alpha\Delta T\delta_{\alpha\beta}$$

$$\sigma_{\alpha\beta} = \frac{E}{1+\nu}\left(\varepsilon_{\alpha\beta} + \frac{\nu}{1-\nu}\varepsilon_{rr}\delta_{\alpha\beta}\right) - \frac{E\alpha\Delta T}{1-\nu}\delta_{\alpha\beta}$$

Representative values for density, and elastic constants of isotropic solids

Most of the data in the table below were taken from the excellent introductory text `Engineering Materials,' by M.F. Ashby and D.R.H. Jones, Pergamon Press.

The follow table's units are GPa for Young's modulus E;

Mgm^{-3} for density;

Material	Mass density ñ	Youngs' Modulus E	Poisson Ratio í	Expansion Coef-ficient K^{-1}
Tungsten Carbide (WC)	14-17	450--650	0.22	5×10^{-6}
Silicon Carbide (SiC)	2.5-3.2	450	0.22	4×10^{-6}
Tungsten (W)	13.4	410	0.30	4×10^{-6}
Alumnium (Al)	3.9	390	0.25	7×10^{-6}
Titanium Carbide (TiC)	4.9	380	0.19	13×10^{-6}

Silicon Nitride (SiN)	3.2	320 -270	0.22	3×10^{-6}
Nickel (Ni)	8.9	215	0.31	14×10^{-6}
CFRP	1.5-1.6	70--200	0.20	2×10^{-6}
Iron (Fe)	7.9	196	0.30	13×10^{-6}
Low alloy steels	7.8	200-210	0.30	15×10^{-6}
Stainless steel	7.5-7.7	190-200	0.30	11×10^{-6}
Mild steel	7.8	196	0.30	15×10^{-6}
Copper (Cu)	8.9	124	0.34	16×10^{-6}
Titanium (Ti)	4.5	116	0.30	9×10^{-6}
Silicon (Si)	2.5-3.2	107	0.22	5×10^{-6}
Silica glass	2.6	94	0.16	0.5×10^{-6}
Aluminum & alloys	2.6-2.9	69-79	0.35	22×10^{-6}
Concrete	2.4-2.5	45-50	0.3	10×10^{-6}
GFRP	1.4-2.2	7-45		10×10^{-6}
Wood	0.4-0.8	9-16	0.2	40×10^{-6}
Polyimides	1.4	3-5	0.1-0.45	40×10^{-6}
Nylon	1.1-1.2	2-4	0.25	81×10^{-6}
PMMA	1.2	3.4	0.35-0.4	50×10^{-6}
Polycarbonate	1.2-1.3	2.6	0.36	65×10^{-6}
Rubbers	0.83-0.91	0.01-0.1	0.49	200×10^{-6}
PVC	1.3-1.6	0.003-0.01	0.41	70×10^{-6}

Other important constants

Bulk, shear and lame modulus are also very important constants to define a material besides Young's modulus and Poisson ratio. Although these two parameters are the most common constant to justify some elastic solids. The concept of Bulk, shear and lame modulus are given below:

$$\text{Bulk Modulus } K = \frac{E}{3(1-2v)}$$

$$\text{Shear Modulus } \mu = \frac{E}{2(1+v)}$$

$$\text{Lame Modulus } \lambda = \frac{vE}{(1+v)(1-2v)}$$

A table about how these important constants related with each other is shown below. The relation of them can be clearly read by this table.

	LAME MODULUS λ	SHEAR MODULUS μ	YOUNG'S MODULUS E	POIS-SON'S RATIO v	BULK MODU-LUS K
λ,μ			$\frac{\mu(3\lambda+2\mu)}{\lambda+\mu}$	$\frac{\lambda}{2(\lambda+\mu)}$	$\frac{3\lambda+2\mu}{3}$
λ,E		Irrational		Irrational	Irrational
λ,v		$\frac{\lambda(1-2v)}{2v}$	$\frac{\lambda(1+v)(1-2v)}{v}$		$\frac{\lambda(1+v)}{2v}$
λ,E		$\frac{3(K-\lambda)}{2}$	$\frac{9K(K-\lambda)}{3K-\lambda}$	$\frac{\lambda}{3K-\lambda}$	
μ,E	$\frac{\mu(2\mu-E)}{E-3\mu}$			$\frac{E-2\mu}{2\mu}$	$\frac{\mu E}{3(3\mu-E)}$
μ,v	$\frac{2\mu v}{1-2v}$		$2\mu(1+v)$		$\frac{2\mu(1+v)}{3(1-2v)}$
μ,K	$\frac{3K-2\mu}{3}$		$\frac{9K\mu}{3K+\mu}$	$\frac{3K-2\mu}{2(3K+\mu)}$	
E,v	$\frac{vE}{(1+v)(1-2v)}$	$\frac{E}{2(1+v)}$			$\frac{E}{3(1-2v)}$
E,K	$\frac{3K(3K-E)}{9K-E}$	$\frac{3EK}{9K-E}$		$\frac{3K-E}{6K}$	

V,K	$\frac{3Kv}{(1+v)}$	$\frac{3K(1-2v)}{2(1+v)}$	$3K(1-2v)$		

Meaning of the constants

It is meaningful to understand the physical interpretation of the important constants-Young's modulus E amd Possion ratio ν.

E-In uniaxial stretch test, the responding stress strain data can form a curved figure. Young's modulus can also be interpreted as the gradient of this figure. Its unit is consist with stress which are both in N/m^2. E also represents the stiffness of the solid material.For example steel's Young's modulus is pretty large with $2.1\times10^{11}\,Pa$, which indicates steel's stiffness is large.

í -Poisson's ratio does not have unit. It is a ratio between strains in different directions. The value of ν usually varies between 0.2 to 0.49. Here is physical meaning of different í values.

- Most material has í to be 0.3.
- í is between -1 to 0.5 for stable materials;
- This material can not be compressed to deform ifí =0.5
- This material's lateral part can not be deformed ifí $=0$
- Ifí <0, the material increase its size instead of decrease, i.e. bar material

á -Thermal expansion coefficient,when heated while press the material's volume will change, this parameter shows the change of such volume. The value of it is usually very small. i.e. Stell has its thermal expansion coefficient as $6-10\times10^{-6}\,K^{-1}$.

Bulk modulus-indicates the material's ability to resist volume changes. Bulk modulus always larger than Elastic modulus.

Shear modulus-indicates the material's ability to resist volume changes while keep the shear from deforming. Shear modulus always smaller than Elastic modulus.

Strain Energy Density

In the strain energy density, there are a few things have been observed:

- There is an interaction between you and the material you want to change, if you want to change the shape of the sample, you need to apply some loads on it;
- For the axial tensile test, the process divides into two stage, stage one when you apply load on the samples the strain energy which is recoverable is stored on the sample, while in stage two unload process, the energy is released . Eventually, it will reach its original shape.
- The final state of the solid sample determines how much work need to be done to deform the sample. The before loading condition is irrelative with it.

Here we gives the definition of **strain energy density**, the energy needed per unit volume to change a material sample from original position to a tension state.

The expression of it is shown below, the strain energy can be divided into two parts.

$$\varepsilon_{ij} = \varepsilon^{\beta}_{ij} + \varepsilon^{T}_{ij}$$

For isotropic material samples,

$$\varepsilon^{T}_{ij} = \alpha \Delta T \delta_{ij}$$

this part only caused by heat caused expansion, and the other part:

$$\varepsilon^{T}_{i_j} = \frac{1+v}{E}\sigma_{ij} - \frac{v}{E}\sigma_{kk}\delta_{ij}$$

This part is caused by load, which is also called elastic/mechanical strain.

The strain energy density can be shown by the follow equation. It shows how much load applied just for elastic loading.

$$U = \frac{1}{2}\sigma_{ij}\varepsilon^{\beta}_{ij}$$

Another expression of it can be:

$$U = \frac{1+v}{2E}\sigma_{ij}\sigma_{ij} - \frac{v}{2E}\sigma_{kk}\sigma_{ij}$$

$$U = \frac{E}{2(1+v)}\varepsilon_{ij}^{\beta}\varepsilon_{ij}^{\beta} + \frac{Ev}{2(1+v)(1-2v)}\varepsilon_{ij}^{\beta}\varepsilon_{ij}^{\beta}$$

In this form,

$$\varepsilon_{ig}^{\beta} = \frac{\partial U}{\partial \sigma_{ij}} \qquad \sigma_{ij} = \frac{\partial U}{\partial \varepsilon_{ij}^{\beta}}$$

ELASTIC STIFFNESS AND COMPLIANCE TENSOR

Many mathematic models, which only works for isotropic material does not work any longer for some materials. Although they deform elastically, the results still not very accurate. The reason lies on the characteristic orientation. Taking a piece of wood for example, the sample's grain orientations is focus on one direction. If apply the load in the same direction, the Young's modulus value is larger than other directions. Fiber composite materials also possess the similar property. From other aspect, single crystal material usually be anisotropic, i.e. microelectronic circuits. Some polycrystalline metals may not be isotropic, i.e. manufacture requirement.

Stress strain relationship which can be used for anisotropic materials should be concluded.

So, it is concluded in the following form:

$$\sigma_{ij} = C_{ijkl}\left(\varepsilon_{kl} - \alpha_{kl}\Delta T\right)$$

C_{ijkl} -Elastic stiffness tensor, it is a fourth order tensor;

$\alpha_{kl} = \alpha_{lk}$ -Thermal expansion coefficient tensor

This relation is not vertible.

$$\varepsilon_{ij} = S_{i\,jkl}\sigma_{kl} + \alpha_{ij}\Delta T$$

$S_{i\,jkl}$ -Elastic compliance tensor

Although the one term of the tensor above has eighty-one components. Some of the term, like C_{ijld} have symmetries below. So you do not have to calculate the eight-one material properties. It greatly makes calculation easier.

$$C_{ijkl} = C_{klij} = C_{jikl} = C_{ijlk}$$

The simplified form of the equation reduced the the complexity to twenty-one. Another tensor, called compliance tensor also have symmetries.

How to understand the symmetries of these tensor, the following things need to be noted:

If a SED (short for strain energy density) have its physical meaning , this elastic stiffness tensor have to be symmetric;

If you want to keep the stress tensor to be symmetric, the elastic stiffness tensor also need to be symmetric;

From these two symmetries, it is not hard to conclude that the same vale of the four tensors.

$$C_{ijkl} = C_{jikl} = C_{klji} \; and \; C_{ijkl} = C_{lkij}, so \; C_{ijkl} = C_{klij} = C_{jikl} = C_{ijlk}$$

A definition need to be clarified before using $C_{ijkl} = C_{klij}$

$$C_{ijkl} = \frac{\partial \sigma_{ij}}{\partial \varepsilon_{kl}}$$

While the stress is the gradient of the SED which mentioned before.

$$\sigma_{ij} = \frac{\partial U}{\partial \varepsilon_{ij}}$$

Thus, we get

$$C_{ijkl} = \frac{\partial^2 U}{\partial \varepsilon_{ij} \partial \varepsilon_{kl}}$$

Another thing need to be illustrated

$$\frac{\partial^2 U}{\partial \varepsilon_{ij} \partial \varepsilon_{kl}} = \frac{\partial^2 U}{\partial \varepsilon_{kl} \partial \varepsilon_{ij}}$$

Thus

$$C_{ijkl} = C_{klij}$$

Finally, a simplifier matrix can be described by using the stress strain relationship.

$$\sigma = C\left(\varepsilon - \alpha \Delta T\right)$$

$$\sigma \begin{bmatrix} \sigma_{11} \\ \sigma_{22} \\ \sigma_{33} \\ \sigma_{23} \\ \sigma_{13} \\ \sigma_{12} \end{bmatrix} \quad c = \begin{bmatrix} c_{11} & c_{12} & c_{13} & c_{14} & c_{15} & c_{16} \\ c_{12} & c_{22} & c_{23} & c_{24} & c_{25} & c_{26} \\ c_{13} & c_{23} & c_{33} & c_{34} & c_{35} & c_{36} \\ c_{14} & c_{24} & c_{34} & c_{44} & c_{45} & c_{46} \\ c_{15} & c_{25} & c_{35} & c_{45} & c_{55} & c_{56} \\ c_{16} & c_{26} & c_{36} & c_{46} & c_{56} & c_{66} \end{bmatrix} \quad \varepsilon = \begin{bmatrix} \varepsilon_{11} \\ \varepsilon_{22} \\ \varepsilon_{33} \\ 2\varepsilon_{13} \\ 2\varepsilon_{12} \end{bmatrix} \quad \alpha = \begin{bmatrix} \alpha_{11} \\ \alpha_{22} \\ \alpha_{33} \\ 2\alpha_{23} \\ 2\alpha_{13} \\ 2\alpha_{12} \end{bmatrix}$$

It should be noticed that $c_{11} \equiv c_{1111}\, \varepsilon_{12} \equiv C_{1122} = C_{2211}$.

It also could be written as

$$\varepsilon = S\sigma + \alpha \Delta T$$

$$S = \begin{bmatrix} s_{11} & s_{12} & s_{13} & s_{14} & s_{15} & s_{16} \\ s_{12} & s_{22} & s_{23} & s_{24} & s_{25} & s_{26} \\ s_{13} & s_{23} & s_{33} & s_{34} & s_{35} & s_{36} \\ s_{14} & s_{24} & s_{34} & s_{44} & s_{45} & s_{46} \\ s_{15} & s_{25} & s_{35} & s_{45} & s_{55} & s_{56} \\ s_{16} & s_{26} & s_{36} & s_{46} & s_{56} & s_{66} \end{bmatrix}$$

Here $s_{11} = S_{1111}, s_{12} = S_{1122}$,

All the eigenvalues of the stiffness and compliance matrices should be non-zero.

Note: The elastic and compliance matrices do not contain shear strain and stress everytime. The reason for including such component is to make calculation convenient even the sample is anisotropic. The order of the material property data should be in right order.

Anisotropic Material

If work on the anisotropic material specimen . The uniaxial load not only introduce extensional but shear deformation. The description of S is better than å .

The description of S is shown below.

$$\varepsilon_{11} = s_{11}\sigma_{11},\ \varepsilon_{22} = s_{12}\sigma_{11},\ \varepsilon_{33} = s_{13}\sigma_{11}$$

$$\varepsilon_{23} = s_{14}\sigma_{11},\ \varepsilon_{13} = s_{15}\sigma_{11},\ \varepsilon_{12} = s_{16}\sigma_{11}$$

The unit of above terms are all Pa,

The description of S is 1/E, uniaxial compliance , can also refer as another term of Poisson's ratio like S_{12}/S_{11}, S_{13}/S_{11};

Shear terms is added to this equation comparing with isotropic samples.

Anisotropic Material's Strain Energy Density

The SED can be expressed in the following equations for samples which is not isotropic.

$$U = \frac{1}{2}C_{ijld}\left(\varepsilon_{ij} - \alpha_{ij}\Delta T\right)\left(\varepsilon_{kl} - \alpha_{kl}\Delta T\right)$$

$$= \frac{1}{2}S_{ijkl}\sigma_{ij}\sigma_{kl}$$

The Difference of Equations between Isotropic and Anisotropic Material.

In the mathematic models, some constants like C_{ij} and S_{ij} is set by special coordinate system, which has axis parallel with the material's symmetry plane in isotropic specimen. If the specimen is not isotropic, the transformation of the coordinate system is required. This transformation is connected with the solid boundary. This change is a little complex for a fourth tensor like C_{ijkl}.

$\{e_1, e_2, e_3\}$ is set to be the tensor base of the stiffness tensor. Another tensor base, which is desired in calculation is $\{m_1, m_2, m_3\}$. To change the base, a transform is needed to accomplish the calculation. So a transformation tensor $\grave{U}_{ij}$ is introduced here. The definition is given by $\grave{U}_{ij} = m_i \bullet e_j$. Another expression by matrix is shown below:

$$[\Omega] = \begin{bmatrix} m_1.e_1 & m_1.e_2 & m_1.e_3 \\ m_2.e_1 & m_2.e_2 & m_2.e_3 \\ m_3.e_1 & m_3.e_2 & m_3.e_3 \end{bmatrix}$$

For this matrix, $\Omega\Omega^T = \Omega^T\Omega = I$. This orthogonal matrix also could be calculated by the other terms, base vectors' angles. So it can be described by the follow equations;

$$\Omega_{ij}^{(m)} = \Omega_{ik}\sigma_{kl}^{(e)}\Omega_{jl}\varepsilon_{ij}^{(m)} = \Omega_{ik}\varepsilon_{kl}^{(e)}\Omega_{jl}\alpha_{ij}^{(m)} = \Omega_{ik}\alpha_{ik}^{(m)} = \Omega_{ik}\alpha_{kl}^{(e)}\Omega_{jl}$$

$$C_{ijkl}^{(m)} = \Omega_{ip}\Omega_{jq}C_{pqrs}^{(e)}\Omega_{kr}\Omega_{ls}$$

Here,

$\sigma_{ij}^{(m)}$ -Stress tensors

$\varepsilon_{ij}^{(m)}$ -Strain tensors

$\alpha_{ij}^{(m)}$ -thermal expansion tensors

$C_{ij}^{(m)}$ -elasticity tensors

Another form for $C_{ij}^{(m)}$ is

$$C^{(m)} = KC^{(e)}K^T$$

The definition of K is

$$K = \begin{bmatrix} K^{(1)}\, 2K^{(2)} \\ K^{(3)}\;\; K^{(4)} \end{bmatrix}$$

$$\left.\begin{aligned} K_{ij}^{(1)} &= \Omega_{ij}^{2} \;\; K_{ij}^{2} = \Omega_{i\,\mathrm{mod}(j+1,3)}\Omega_{i\,\mathrm{mod}(j+2,3)} \\ K_{ij}^{(3)} &= \Omega_{\mathrm{mod}(i+1,3)j}\Omega_{\mathrm{mod}(i+2,3)} \\ K_{ij}^{(4)} &= \Omega_{\mathrm{mod}(i+1,3)\mathrm{mod}(j+1,3)}\Omega_{\mathrm{mod}(i+2,3)\mathrm{mod}(j+2,3)} \\ &\quad + \Omega_{\mathrm{mod}(i+1,3)\mathrm{mod}(j+2,3)}\Omega_{\mathrm{mod}(i+2,3)\mathrm{mod}(j+1,3)} \end{aligned}\right\} i, j = 1, 3$$

Here , modulus is a function, i.e.

$$\mathrm{mod}(i,3) = \begin{cases} i & i \le 3 \\ i - 3 i > 3 \end{cases}$$

This descriptions of these terms make it easier for computational calculation.

S could also be in the following form:

$$S^{(m)} = K^{-T} S^{(e)} K^{-1}$$

here

$$K^{-T} = \begin{bmatrix} K^{(1)} & K^{(2)} \\ 2K^{(3)} & K^{(4)} \end{bmatrix}$$

The detail of this form can be found in the book *Anisotropic Elasticity: Theory and Applications, 1996.* It will not be repeated here.

If the rotating angle of the base is in a counterclockwise sense, the original form could be simplified to the follow.

$$\begin{bmatrix} 1 & 0 & 0 & 0 & 0 & 0 \\ 0 & c^2 & s^2 & 2cs & 0 & 0 \\ 0 & s^2 & c^2 & -2cs & 0 & 0 \\ 0 & -cs & cs & c^2-s^2 & 0 & 0 \\ 0 & 0 & 0 & 0 & c & -s \\ 0 & 0 & 0 & 0 & s & c \end{bmatrix} \begin{bmatrix} c^2 & 0 & s^2 & 0 & 2cs & 0 \\ 0 & 1 & 0 & 0 & 0 & 0 \\ s^2 & 0 & c^2 & 0 & -2cs & 0 \\ 0 & 0 & 0 & c & 0 & -s \\ -cs & 0 & cs & 0 & c^2-s^2 & 0 \\ 0 & 0 & 0 & s & 0 & c \end{bmatrix} \begin{bmatrix} c^2 & s^2 & 0 & 0 & 0 & 2cs \\ s^2 & c^2 & 0 & 0 & 0 & -2cs \\ 0 & 0 & 1 & 0 & 0 & 0 \\ 0 & 0 & 0 & c & s & 0 \\ 0 & 0 & 0 & -s & c & 0 \\ -cs & cs & 0 & 0 & 0 & c^2 & s^2 \end{bmatrix}$$

$$c = \cos\theta$$

$$s = \sin$$

K^{-1} -it is easy to obtain by inversing the sign

Note: Orientation change may occur if successively rotation is applied.

Major compenent of the material, like $\sigma - \varepsilon$,C and a are independent of base variation.

Symmetry in Anisotropic Materials

A general anisotropic solid has 21 independent elastic constants. Note that in general, tensile stress may induce shear strain, and shear stress may cause extension.

What's more, axial tensile stretch also effects shear, which turn on to effect extension behavior.

When the solid possesses symmetric plane, if conduct load particular or align of this plane, The deformation only occur in such direction.

In the case of a solid material, if it have one symmetry plane and basic component e1 particular with this plane, the stiffness matrix should be like this:

$$c_{15} = c_{16} = c_{25} = c_{26} = c_{35} = c_{36} = 0$$

The same as the symmetrical terms, which

$$c_{1112} = c_{1113} = c_{2212} = c_{2213} = c_{3312} = c_{3313} = 0.$$

It leads to the reduce of the matrix to thirteen terms.

So the symmetry of the solid materials could be helpful simplify the matrix dimension. You can prove it yourself if interested.

The stress strain relationship between other type of solid materials will be presented in the next few parts.

Orthotropic Materials

Orthotropic solid material possesses three symmetry planes. For this kind of material, there are nine terms on the stress-strain relation and all the base vectors are normal to the plane, stiffness matrix has the flowing expression:

$$C = \begin{bmatrix} c_{11} & c_{12} & c_{13} & 0 & 0 & 0 \\ & c_{22} & c_{23} & 0 & 0 & 0 \\ & & c_{33} & 0 & 0 & 0 \\ & sym & & c_{44} & 0 & 0 \\ & & & 0 & c_{55} & 0 \\ & & & 0 & 0 & c66 \end{bmatrix}$$

Another expression of this matrix is showing below, which in term of Young's modulus and Poisson ratio. These terms are generalized, they have similar meaning with the definition we described before.

$$S = \begin{bmatrix} 1/E_1 & -v_{21}/E_2 & -v_{31}/E_3 & 0 & 0 & 0 \\ -v_{12}/E_1 & 1/E_2 & -v_{32}/E_3 & 0 & 0 & 0 \\ -v_{13}/E_1 & -v_{23}/E_2 & 1/E_3 & 0 & 0 & 0 \\ 0 & 0 & 0 & 1/\mu_{23} & 0 & 0 \\ 0 & 0 & 0 & 0 & 1/\mu_{23} & 0 \\ 0 & 0 & 0 & 0 & 0 & 1/\mu_{12} \end{bmatrix}$$

here

$$v_{ij}/E_i = v_{ji}/E_j$$

The poisson ratio is asymmetric, so this definition makes the matrix to be symmetric.

The compliance term is shown below, they are often used as engineering term.

$$c_{11} = E_1\left(1-\nu_{23}\nu_{32}\right)\gamma \quad c_{22} = E_2\left(1-\nu_{13}\nu_{31}\right)\gamma \quad c_{33} = E_3\left(1-\nu_{12}\nu_{21}\right)\gamma$$

$$c_{12} = E_1\left(\nu_{21}+\nu_{31}\nu_{23}\right)\gamma = E_2\left(\nu_{12}+\nu_{32}\nu_{13}\right)\gamma$$

$$c_{13} = E_1\left(\nu_{31}+\nu_{21}\nu_{32}\right)\gamma = E_3\left(\nu_{13}+\nu_{12}\nu_{23}\right)\gamma$$

$$c_{23} = E_2\left(\nu_{32}+\nu_{12}\nu_{31}\right)\gamma = E_3\left(\nu_{23}+\nu_{21}\nu_{13}\right)\gamma$$

$$c_{44} = \mu_{23} \quad c_{55} = \mu_{13} \quad c_{66} = \mu_{12}$$

$$\gamma \frac{1}{1-\nu_{12}\nu_{21}-\nu_{23}\nu_{32}-\nu_{31}\nu_{13}-2\nu_{21}\nu_{32}\nu_{13}}$$

Another expression is

$$E_1 = \left(c_{11}c_{22}c_{23}+2c_{23}c_{12}c_{13}-c_{11}c_{23}^2-c_{22}c_{13}^2-c_{33}c_{12}^2\right)/\left(c_{22}c_{33}-c_{23}^2\right)$$

$$E_2 = \left(c_{11}c_{22}c_{33}+2c_{23}c_{12}c_{13}-c_{11}c_{23}^2-c_{22}c_{13}^2-c_{33}c_{12}^2\right)/\left(c_{11}c_{33}-c_{13}^2\right)$$

$$E_3 = \left(c_{11}c_{22}c_{33}+2c_{23}c_{12}c_{13}-c_{11}c_{13}^2-c_{22}c_{13}^2-c_{33}c_{12}^2\right)/\left(c_{11}c_{22}-c_{12}^2\right)$$

$$\nu_{21} = \left(c_{12}c_{33}-c_{13}c_{23}\right)/\left(c_{11}c_{33}-c_{13}^2\right) \qquad \nu_{12} = \left(c_{12}c_{33}-c_{13}c_{23}\right)/\left(c_{22}c_{33}-c_{23}^2\right)$$

$$\nu_{31} = \left(c_{13}c_{22}-c_{12}c_{23}\right)/\left(c_{11}c_{22}-c_{12}^2\right) \qquad \nu_{13} = \left(c_{22}c_{13}-c_{12}c_{23}\right)/\left(c_{22}c_{33}-c_{23}^2\right)$$

$$\nu_{23} = \left(c_{11}c_{23}-c_{12}c_{13}\right)/\left(c_{11}c_{33}-c_{13}^2\right) \qquad \nu_{32} = \left(c_{11}c_{23}-c_{12}c_{13}\right)/\left(c_{11}c_{22}-c_{12}^2\right)$$

$$\mu_{23} = c_{44}, \quad \mu_{13} = c_{55} \quad \mu_{12} = c_{66}$$

If an orthotropic solid sample is expanding by heat, this deformation will not cause shearing deformation .But the deformation caused by heat required to be same value in all the three directions. So the thermal expansion coefficient tensor is expressed by the below description.

$$\begin{bmatrix} \alpha_1 & 0 & 0 \\ 0 & \alpha_2 & 0 \\ 0 & 0 & \alpha_3 \end{bmatrix}$$

Transversely Isotropic Material

If an orthotropic solid includes an isotropic plane. It indicates the axial direction rotation will not cause the change of the solid's state.

For a transversely isotropic solid material,

$$c_{22} = c_{11}\, c_{23} = c_{13}, c_{55} = c_{44}, c_{66} = (c_{11} - c_{12})/2,$$

The stiffness matrix can be expressed in the follow form

$$C = \begin{bmatrix} c_{11} & c_{12} & c_{13} & 0 & 0 & 0 \\ & c_{11} & c_{13} & 0 & 0 & 0 \\ & & c_{33} & 0 & 0 & 0 \\ & & sym & c_{44} & 0 & 0 \\ & & & 0 & c_{44} & 0 \\ & & & 0 & 0 & (c_{11} - c_{12})/2 \end{bmatrix}$$

e_3 is the basic vector which is norm to the symmetric plane. This constant has such requirement.

$$E_1 = E_2 = E_p \quad E_3 = E_t$$

$$v_{12} = v_{21} = v_p \quad v_{31} = v_{32} = v_{tp}$$

$$v_{13} = v_{23} = v_{pt}$$

Compliance matrix has the following expression.

$$S = \begin{bmatrix} 1/E_1 & -v_{21}/E_2 & -v_{31}/E_3 & 0 & 0 & 0 \\ -v_{12}/E_1 & 1/E_2 & -v_{32}/E_3 & 0 & 0 & 0 \\ -v_{13}/E_1 & -v_{23}/E_2 & 1/E_3 & 0 & 0 & 0 \\ 0 & 0 & 0 & 1/\mu_t & 0 & 0 \\ 0 & 0 & 0 & 0 & 1/\mu_t & 0 \\ 0 & 0 & 0 & 0 & 0 & 1/\mu_p \end{bmatrix}$$

In the above expression,

$$\mu_p = E_p / 2(1 - v_p)$$

$$v_{tp} / E_t = v_{pt} / E_p$$

For the poisson ratio is asymmetric.

The following conditions should also be considered.

$$c_{11} = c_{22} = E_p\left(1 - v_{pt}v_{tp}\right)\gamma \;\; c_{33} = E_t\left(1 - v_p^2\right)\gamma \;\; c_{12} = E_p\left(v_p + v_{pt}v_{tp}\right)\gamma$$

$$c_{13} = c_{23} = E_p\left(v_{tp} + v_p v_{tp}\right)\gamma = E_t\left(v_{pt} + v_p v_{pt}\right)\gamma \;\; c_{44} = \mu_t \;\; c_{66} = \mu_p$$

$$\gamma \frac{1}{1 - v_p^2 - 2v_{pt}v_{tp} - v_{tp}}$$

$$E_p = \left(c_{11}^2 c_{33} + 2c_{13}^2 c_{12} - 2c_{11}c_{13}^2 - c_{33}c_{12}^2\right) / \left(c_{11}c_{33} - c_{13}^2\right)$$

$$E_t = \left(c_{11}^2 c_{33} + 2c_{13}^2 c_{12} - 2c_{11}c_{13}^2 - c_{33}c_{12}^2\right) / \left(c_{11}^2 - c_{12}^2\right)$$

$$v_p = \left(c_{12}c_{33} - c_{13}^2\right) / \left(c_{11}c_{33} - c_{13}^2\right), \;\; v_{tp} = \left(c_{13}c_{11} - c_{12}c_{13}\right) / \left(c_{11}^2 - c_{12}^2\right)$$

$$v_{pt} = \left(c_{11}c_{13} - c_{12}c_{13}\right) / \left(c_{11}c_{33} - c_{13}^2\right), \;\; \mu_{23} = c_{44}, \;\; \mu_{13} = c_{55} \;\; \mu_{12} = c_{66}$$

The thermal expansion coefficient tensor is expressed by the below description, since it in the symmetric plane need to be same.

$$\begin{bmatrix} \alpha_1 & 0 & 0 \\ 0 & \alpha_2 & 0 \\ 0 & 0 & \alpha_3 \end{bmatrix}$$

Hexagonal Close Packed (HCP) Structured Transversely Isotropic Material's Elastic conStants

Here is an example of transversely isotropic material, HCP crystals.

In hexagonal close-packed crystals, e_3 is normal to the 0001 plane , which is basal plane. e_1 and e_2 are perpendicular to symmetric plane , the direction of them can be random.

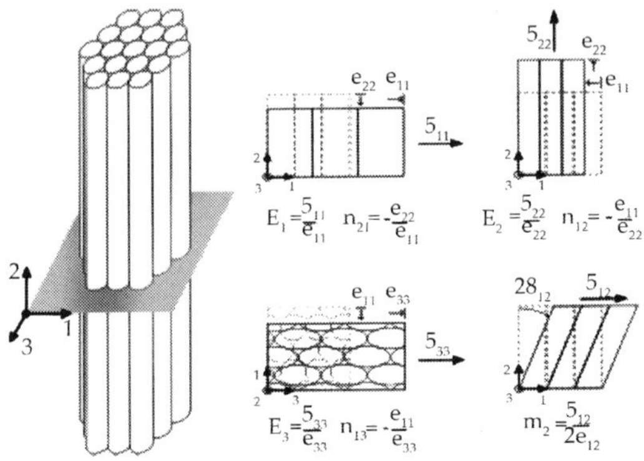

Normal HPC metal sample's stuffiness is shown below.

	C_{11} (GPa)	C_{33} (GPa)	C_{44} (GPa)	C_{13} (GPa)	C_{13} (GPa)
Be	292.3	336.4	162.5	26.7	14
C	1160	46.6	2.3	290	109
Cd	115.8	51.4	20.4	39.8	40.6
Co	307	358.1	78.3	165	103
Hf	181.1	196.9	55.7	77.2	66.1
Mg	59.7	61.7	16.4	26.2	21.7
Ti	162.4	180.7	46.7	92	69
Zn	161	61	38.3	34.2	50.1
Zr	143.4	164.8	32	72.8	65.3
ZnO	209.7	210.9	42.5	121.1	105.1

All the important constants can be obtained by the definitions.

	E_p (GPa)	Et (GPa)	Vp	vtp	vpt	µt (GPa)	µp (GPa)
Be	289.38	335.17	0.09	0.04	0.04	162.50	132.80
C	903.69	30.21	0.04	0.08	2.25	2.30	435.00
Cd	83.02	30.21	0.09	0.26	0.72	20.40	38.00
Co	211.30	313.15	0.49	0.22	0.15	78.30	71.00
Hf	139.87	163.07	0.35	0.26	0.22	55.70	51.95
Mg	45.45	50.74	0.36	0.25	0.23	16.40	16.75

Ti	104.37	143.27	0.48	0.27	0.20	46.70	35.20
Zn	119.45	35.28	0.06	0.26	0.87	38.30	63.40
Zr	98.79	125.35	0.40	0.30	0.24	32.00	35.30
ZnO	127.30	144.12	0.44	0.32	0.28	42.50	44.30

Cubic Structured Materials' Elastic Property

Face Centered Cubic (FCC) and Body Centered Cubic (BCC) materials are symmetric. So the constitutive equation for this kind of material is direct and simple. The term of their material is only three. The tensor base is very easy to be choose from.

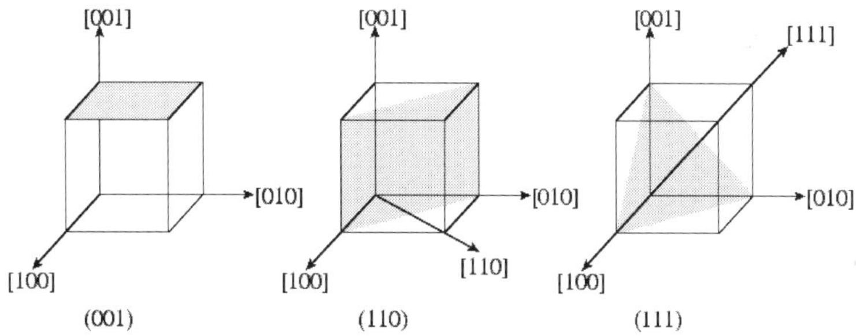

So

$$\begin{bmatrix} \sigma_{11} \\ \sigma_{22} \\ \sigma_{33} \\ \sigma_{23} \\ \sigma_{12} \\ \sigma_{13} \end{bmatrix} = \begin{bmatrix} c_{11} & c_{12} & c_{13} & 0 & 0 & 0 \\ & c_{11} & c_{13} & 0 & 0 & 0 \\ & & c_{33} & 0 & 0 & 0 \\ & & sym & c_{44} & 0 & 0 \\ & & & 0 & c_{44} & 0 \\ & & & 0 & 0 & c_{44} \end{bmatrix} \begin{bmatrix} \varepsilon_{11} \\ \varepsilon_{22} \\ \varepsilon_{33} \\ 2\varepsilon_{23} \\ 2\varepsilon_{12} \\ 2\varepsilon_{13} \end{bmatrix}$$

In other form expression,

$$\begin{bmatrix} \varepsilon_{11} \\ \varepsilon_{22} \\ \varepsilon_{33} \\ 2\varepsilon_{23} \\ 2\varepsilon_{12} \\ 2\varepsilon_{13} \end{bmatrix} = \begin{bmatrix} 1/E & -v/E & -v/E & 0 & 0 & 0 \\ -v/E & 1/E & -v/E & 0 & 0 & 0 \\ -v/E & -v/E & 1/E & 0 & 0 & 0 \\ 0 & 0 & 0 & 1/\mu & 0 & 0 \\ 0 & 0 & 0 & 0 & 1/\mu & 0 \\ 0 & 0 & 0 & 0 & 0 & 1/\mu \end{bmatrix} \begin{bmatrix} \sigma_{11} \\ \sigma_{22} \\ \sigma_{33} \\ \sigma_{23} \\ \sigma_{12} \\ \sigma_{13} \end{bmatrix}$$

So cubic structured material is same with normal isotropic solid material. Only μ is independent with Young's modulus and Poisson ratio. The constant A's definition is given below.

$$A = \frac{2\mu(1+v)}{E} = \frac{2c_{44}}{c_{11} - c_{12}}$$

It in fact gives the definition of the how anistropic the material is

For this material the thermal expansion coefficient matrix must be isotropic.

For example, if $\mathrm{A} = 1,$ the sample is isotropic.

Another equation show how these constant related with each other shown below.

$$E = \left(c_{11}^2 + c_{12}c_{11} - 2c_{12}^2\right)/\left(c_{11} + c_{12}\right) \quad v = c_{12}/\left(c_{11} + c_{12}\right) \quad \mu = c_{44}$$

$$c_{11} = E(1-v)/\left(1 - v - 2v^2\right) \quad c_{12} = Ev/\left(1 - v - 2v^2\right)$$

Cubic Materials' Important Constants

A table of elastic constants for various cubic crystals and compounds (modified from Simmons and Wang 'Single Crystal Elastic Constants and Calculated Aggregate Properties' MIT Press (1970)) is given below

The follow table shows the normal cubic material's constants' values.

(This table is from Simmons and Wang's single crystal elastic constants and calculated aggregate properties ,1970

Material		c_{11}(GPa)	c_{44}(GPa)	c_{12}(GPa)	E(GPa)	v	μ(GPa)	A
Ag	(fcc)	124.00	46.10	93.40	43.75	0.43	46.10	3.01
Al	(fcc)	107.30	28.30	60.90	63.20	0.36	28.30	1.22
Au	(fcc)	192.90	41.50	163.80	42.46	0.46	41.50	2.85
Cu	(fcc)	168.40	75.40	121.40	66.69	0.42	75.40	3.21
Ir	(fcc)	580.00	256.00	242.00	437.51	0.29	256.00	1.51
Ni	(fcc)	246.50	127.40	147.30	136.31	0.37	127.40	2.57
Pb	(fcc)	49.50	14.90	42.30	10.52	0.46	14.90	4.14
Pd	(fcc)	227.10	71.70	176.00	73.41	0.44	71.70	2.81
Pt	(fcc)	346.70	76.50	250.70	136.29	0.42	76.50	1.59
Cr	(bcc)	339.80	99.00	58.60	322.56	0.15	99.00	0.70
Fe	(bcc)	231.40	116.40	134.70	132.28	0.37	116.40	2.41
K	(bcc)	4.14	2.63	2.21	2.60	0.35	2.63	2.73
Li	(bcc)	13.50	8.78	11.44	3.00	0.46	8.78	8.52
Mo	(bcc)	440.80	121.70	172.40	343.86	0.28	121.70	0.91
Na	(bcc)	6.15	5.92	4.96	1.72	0.45	5.92	9.95
Nb	(bcc)	240.20	28.20	125.60	153.95	0.34	28.20	0.49
Ta	(bcc)	260.20	82.60	154.50	145.08	0.37	82.60	1.56
V	(bcc)	228.00	42.60	118.70	146.72	0.34	42.60	0.78
W	(bcc)	522.40	160.80	204.40	407.43	0.28	160.80	1.01
C	(dc)	949.00	521.00	151.00	907.54	0.14	521.00	1.31
Ge	(dc)	128.40	66.70	48.20	102.09	0.27	66.70	1.66
Si	(dc)	166.20	79.80	64.40	130.23	0.28	79.80	1.57
GaAs		118.80	59.40	53.70	85.37	0.31	59.40	1.82
GaP		141.20	70.50	62.50	102.85	0.31	70.50	1.79
InP		102.20	46.00	57.60	60.68	0.36	46.00	2.06
KCl		39.50	6.30	4.90	38.42	0.11	6.30	0.36
LiF		114.00	63.60	47.70	85.86	0.29	63.60	1.92
MgO		287.60	151.40	87.40	246.86	0.23	151.40	1.51
NaCl		49.60	12.90	12.40	44.64	0.20	12.90	0.69
TiC		500.00	175.00	113.00	458.34	0.18	175.00	0.90

NON-LINEAR SOLID MECHANICS

Non-linear effects cannot be omitted in practical problems. The linear material property shown above cannot solve the real problem precisely. For example, if the sample have an extreme size (either large or small) or complex geometry, it will introduce non-linear behavior.

We will focus on these problems here.

The introduction of non-linear effects can divided into two situation, material and geometric.

For material property, the reason for this non-linearity is from the stress strain relation. The figure of stress and strain is not linear , which can be observed from tensile tests. For example, the elastic plastic behavior.

After deforming, the solid sample has two parts of the sample, the deformed part is different with the undeformed part with finite deformation. Since the strain and displacement relationship changes to be non-linear, the original equations can not be used here due to the unchanged shape. What if the finite deformation happens the unnormal behavior like bending and branching will cause non-linear effects. i.e. Euler column. The bending phenomenon with just axial force.

Before introducing another situation. Some geometry need to describe firstly.

Plate-flat geometry, the small dimension is called thickness

Shell-curved geometry, also have thickness parameters

Beam-Two dimensions

Frame-Two dimensions

Rod-Two dimensions

If the sample has different size in each dimensions, and the difference is extremely large. The solution of such problem will not be very precise if using the 3-D method and all the mathematic models shown before. The major reason is the abnormal conditions makes the algebraic questions very hard to solve and the error for solving such problem is large.

Assumption of Small Deformation

There are so many beautiful equations about the linear situation, how these equations evolution situation if the conditions turn from linear to not linear? This section will illustrate this problem, especially from the stress strain relation.

What constructed the mathematic models for solid problem?

(1) Equilibrium theory

(2) Strain displacement relationship

(3) Material constitutive equations

(4) BC

(5) IC

All the problem will be introduced by two forms:

Cartesian Tensor and Matrix

If we want to depict the large deformation situation with large strain. In order to state the stiffness related with geometry and IC , the first form is a good choice to make this description easier.

If we want to focus on the steps for small deformation, both forms need to be considered, even the transformation between these two forms.

InCartesiansystemthecoordinatesalwaysbenotedby x, y, z or x_1, x_2, x_3 . While the displacements will be noted as u, v, w or u_1, u_2, u_3. It is usually in small deformation problem. In 3D problem they are x_1, x_2, x_3 and u_1, u_2, u_3. While in 2D problem they become x_1, x_2 and u_1, u_2. The follow expression is the coordinates in matrix form:

$$\text{x} = \begin{Bmatrix} x \\ y \\ z \end{Bmatrix} = \begin{Bmatrix} x_1 \\ x_2 \\ x_3 \end{Bmatrix}$$

u is the displacements

$$\text{u} = \begin{Bmatrix} u \\ v \\ w \end{Bmatrix} = \begin{Bmatrix} u_1 \\ u_2 \\ u_3 \end{Bmatrix}$$

In constitutive model all the fields will be mixed together.

To get this model by this method, we use mean and deviatoric effects method, the definition of this operator is shown below.

$$m = \begin{Bmatrix} 1 \\ 1 \\ 1 \\ 0 \\ 0 \\ 0 \end{Bmatrix} ; I_d = I - \frac{1}{3} m m^T$$

I- Identity matrix

Sample stain could by described by the volume change and pressure separately, ε_v is the volume change, the description is shown below,

$$\varepsilon = I_d (Su) + \frac{1}{3} m \varepsilon_v$$

p is the pressure, which is also the mean stress, this description also is given

$$\sigma = I_d \sigma + mp$$

$\hat{\sigma}$ -reduced stress, i.e incremental strains

This terms is relative with material constitutive model. Here, this is given by the follow expression.

$$\hat{\sigma} = \sigma(e)$$

Here, there is no need to consider mean and deviatoric effect.

The equations with transients effects is shown below, it is called Galerkin equations, which we mentioned before.

$$\int_{\Omega} \delta u^T \rho \ddot{u}\, d\Omega + \int_{\Omega} \delta (Su)^T \sigma d\Omega = \int_{\Omega} \delta u^T b d\Omega + \int_{\Gamma_t} \delta u^T t d\Gamma$$

$$\int_{\Omega} \delta \varepsilon_V \left[\frac{1}{3} m^T \sigma - p \right] d\Omega = 0$$

$$\int_{\Omega} \delta p \left[m^T (Su) - \varepsilon_V \right] d\Omega = 0$$

To calculate the gradient of this equation, let,

$$d\sigma = D_T\, d\varepsilon$$

In the mixed form, it turns to be,

$$K_T = \int_{\Omega} \begin{bmatrix} B^T & B_V^T \end{bmatrix} \begin{bmatrix} I_d \\ \frac{1}{3} m^T \end{bmatrix} D_T \begin{bmatrix} I_d & \frac{1}{3} m \end{bmatrix} \begin{bmatrix} B \\ B_V \end{bmatrix} d\Omega$$

One constant term of the equation is defined by the following

$$\bar{D}_T = \begin{bmatrix} I_d \\ \frac{1}{3} m^T \end{bmatrix} D_T \begin{bmatrix} I_d & \frac{1}{3} m \end{bmatrix} = \begin{bmatrix} I_d D_T I_d & \frac{1}{3} I_d D_T m \\ \frac{1}{3} m^T D_T I_d & \frac{1}{9} m^T D_T m \end{bmatrix}$$

Until this step , I_d and m has already made the calculation much easier. Thus, it makes the construction of $\bar{B}$ very easy by multiplying these two terms with B and B_V separately. The description of $\bar{B}$ is,

$$\bar{B} = I_d B + \frac{1}{3} m B_V$$

This gives a general form of equations which can be used in both linear and not linear constitutive models. The strain and stress figure not only showing the stiffness property, but also contain many other material property like plasticity and viscoelasticity. All the equations need to meet one need-Incompressible solid materials.

So there are two forms:

4-noded quadrilateral / 8-noded brick elements ;

9-noded quadrilateral / 27-noded brick elements;

If the problem is in 2d , we use,

$$N_P = N_V = \begin{bmatrix} 1 & \xi & \eta \end{bmatrix} \text{ or} \begin{bmatrix} 1 & x & y \end{bmatrix}$$

If the problem is in 3d , we use

$$N_P = N_V = \begin{bmatrix} 1 & \xi & \eta & \zeta \end{bmatrix} \; or \begin{bmatrix} 1 & x & y & z \end{bmatrix}$$

This is the general method to solve most problems in solid fields.

Quasi-Harmonic Field Problems

We consider a simple quasi-harmonic problem given by

For quasi-harmonic field situation, let's look at one case,

$$\rho c \dot{\phi} + \nabla^{\mathrm{T}} \mathbf{q} - Q(\phi) = 0$$

This equation should together with related BC. This equation could be used in a large varity of industry field. i.e, thermal problem, solid magnetic problem, fluid flow. Q could be defined as the follow,

$$\mathbf{q} = \mathbf{q}(\phi, \nabla\phi) = \mathbf{k}(\phi, \nabla\phi)\nabla\phi$$

Linearize this form to read easily

$$d\mathbf{q} = \mathbf{k}^0 d\phi + \mathbf{k}^1 d(\nabla\phi)$$

In this equation

$$k_i^0 = \frac{\partial q_i}{\partial \phi} \qquad \text{and} \qquad k_{ij}^1 = \frac{\partial q_i}{\partial \phi_{,j}}$$

Q is also a non-linear term regarding as heat source.

Discrete it,

$$\delta\Pi = \int_{\Omega} \delta\phi \rho c \dot{\phi} \, d\Omega - \int_{\Omega} (\nabla \delta\phi)^{T} \mathbf{q} \, d\Omega$$

$$- \int_{\Omega} \delta\phi Q(\phi) \, d\Omega - \int_{\Gamma_q} \delta\phi \bar{q}_n \, d\Gamma = 0$$

This form is still useful if heat source related with $\varnothing$. Here,

$$\phi = \mathbf{N}\tilde{\boldsymbol{\phi}} \quad \text{and} \quad \delta\phi = \mathbf{N}\delta\tilde{\boldsymbol{\phi}}$$

In another form,

$$\boldsymbol{\Psi} = \mathbf{f}(\tilde{\boldsymbol{\phi}}) - \mathbf{C}\dot{\tilde{\boldsymbol{\phi}}} - \mathbf{P}_q(\tilde{\boldsymbol{\phi}}) = \mathbf{0}$$

In this equation

$$\mathbf{C} = \int_{\Omega} \mathbf{N}^{T} \rho c \mathbf{N} \, d\Omega$$

$$\mathbf{P}_q = -\int_{\Omega} \mathbf{N}^{T} \mathbf{q} \, d\Omega$$

$$\mathbf{f} = \int_{\Omega} \mathbf{N}^{T} Q(\phi) \, d\Omega - \int_{\Gamma_q} \mathbf{N}^{T} \bar{q}_n \, d\Gamma$$

Once again we have the choice of using $\varnothing_{n+1}$ or $\dot{\varnothing}_{n+1}$ as the primary solution variable. To this extent the process of solving transient problems follows absolutely the same lines as those described in the previous contents.

Primary solution need to be chosed here: $\varnothing_{n+1}$ and $\dot{\varnothing}_{n+1}$.

Examples of Non-Linear Problem

Transient Heat Conduction

As mentioned before about the non-linear problem solution. Set $\varnothing$ to be the current temperature.

The non-linearity can vary depending on specific heat, c, thermal conductivity, k and source Q, temperature dependence, or radiation boundary conditions

$$k\frac{\partial T}{\partial n} = -\alpha(T - T_a)^{n}$$

N $\neq$ 1. Where α is the convective transfer coefficient and T_a is the outside temperature. Two examples will be used to illustrate this.

The first problem is to freeze the ground, where the latent heat is expressed by changing the material properties of the temperature in a narrow area, as shown in Figure 4.1. In addition, in the transition from the fluid to the frozen state, the conductivity changes. Thus, we now have a problem that the matrix C and P are variable, and the solution in Figure 4.2 illustrates the progress of the frozen frontier derived by using the three-point algorithm of $C = C_n$ and $P = P_n$

There are some computational features in this problem because the value of the specific heat in the transition zone becomes very high and if the temperature step crosses the freezing point, the time step can be missed. In order to avoid this difficulty, keep the heat balance correct and introduce the concept of enthalpy to express

$$H = \int_0^T \rho c dT$$

At this time, whenever the temperature change is taken into account, the responsible ρc value is calculated and the correct H is given

The heat transfer problem of phase change is very important in welding and casting techniques. There are some very useful finite element solutions.

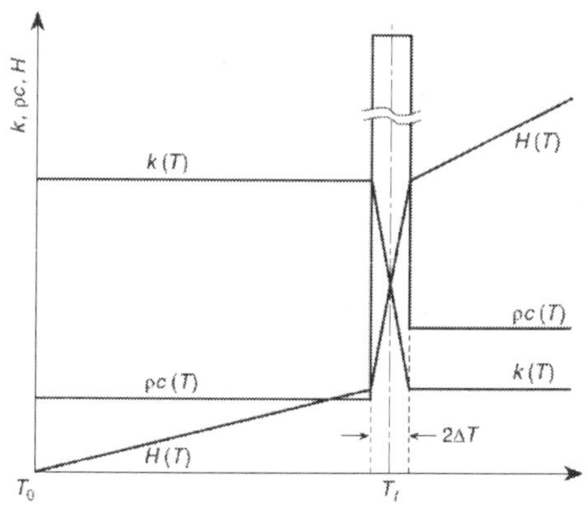

Figure 4.1 the thermal physical properties are estimated to vary during phase

change. The latent heat effect is approximated by a large capacity over a small temperature interval $2\Delta T$.

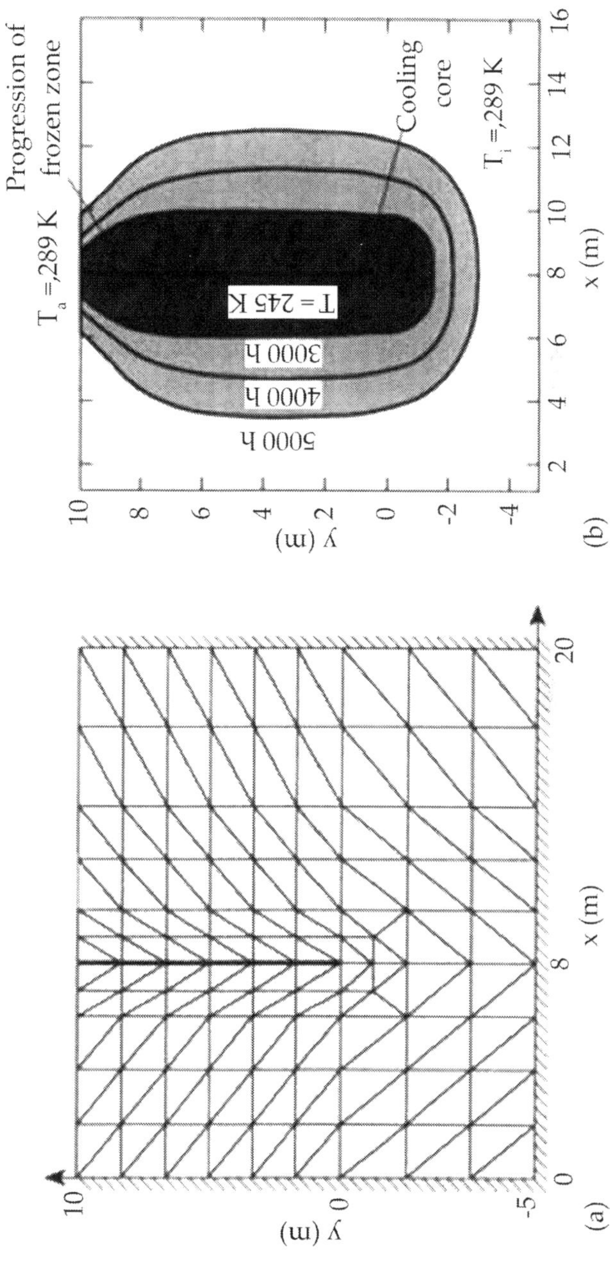

Figure 4.2 Freezing of a moist solil (sand)

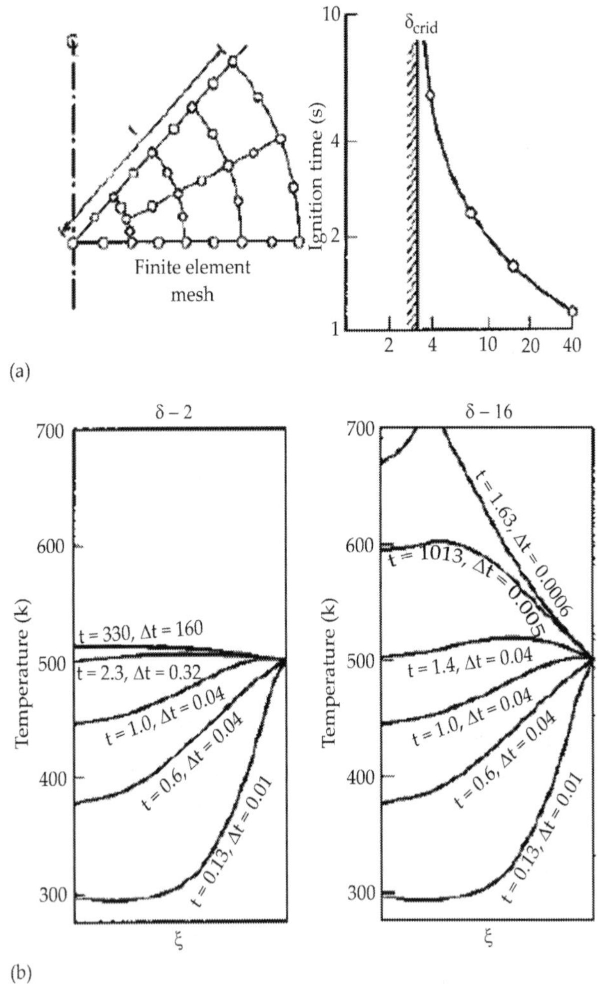

Figure 4.3 Reactive sphere. How prompt tempretaure act for ignition $\left(\bar{\ddot{a}} = 16\right)$ and not ignition $\left(\bar{\ddot{a}} = 2\right)$ a) time profiles versus Frank-Kamenetskii parameter; b) temperature profiles for ignition ($\bar{\ddot{a}} = 16$) and non-ignition $\left(\bar{\ddot{a}} = 2\right)$ transient behavior of a reactive sphere.

The second non-linear example concerns the problem of spontaneous ignition. Here the heat generated depends on the temperature

The second example which is non-linear problem deals with the spontaneous ignition. The heat generated here depends on the temperature as shown below

$$Q = \bar{\delta} e^{T}$$

When the calculated temperature continues to rise to unormal values, the problem may become unstable. In Figure 4.3, it shows a transient solution with an initial value of T = 290K and immersed in a 500 K bath. For two values of the parameter $\bar{\delta}$ for of $k = \rho c = 1$ the linearity is so serious that the iteration is required for each time increment. For a larger value of the delta value, the temperature increases to an infinite value over a finite time and the calculated time interval must be changed accordingly. The finite time of this point is called the induction time, and the values of $\bar{\delta}$ .for are shown in Figure 4.3.

In the calculation process, there is a problem of changing the time interval, which has not yet been discussed, but this problem must be clearly stated, as it often occurs, in order to avoid inaccurate and need to be described in order to avoid major changes in the results.

Structural Dynamics

In these examples, the transient transitions of materials and geometric nonlinearities affect the results. Resulting in geometrical and material nonlinearity, damping force will be ignored.

There is a technique that is very effective against this problem, and the steady-state condition is set to $\ddot{u} = \ddot{u} \cong 0$, at which point the solution is obtained. And this method can also introduce "dynamic relaxation", in the limited difference to solve the problem. Is a general solution.

The examples shown below are explicit dynamic methods.

In the first question, the board 3 shows a large three-dimensional problem, and its solution is obtained using a explicite dynamic scheme. In this case, the implicite solution will not be available at all, and explicit code provides a very effective way to solve the crash problem. However, it must be recognized that such a final solution is not necessary.

The second example, Figure 4.4 shows a typical crash analysis of a motor vehicle by similar means.

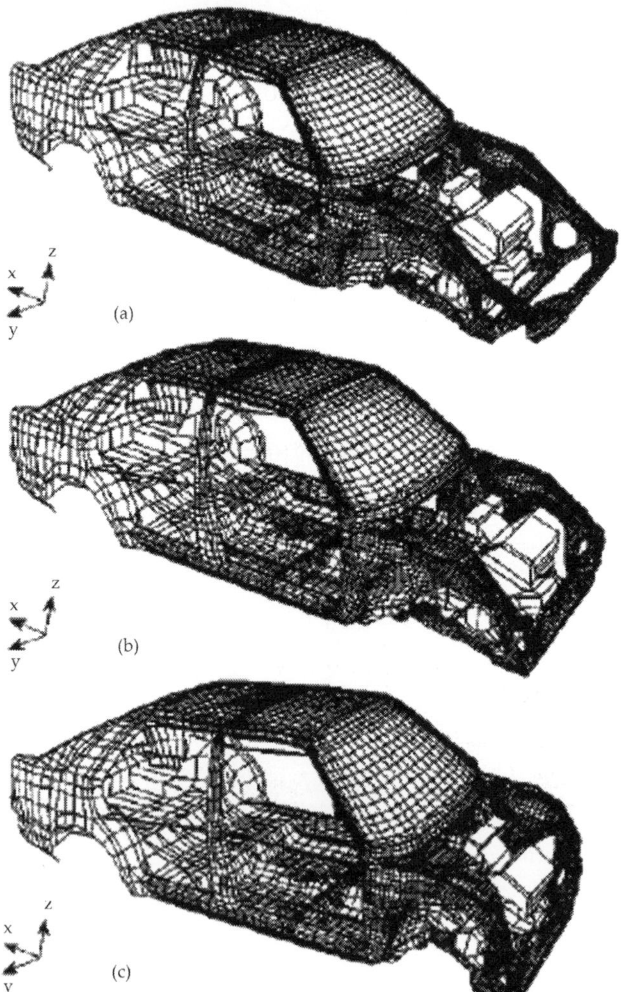

Figure 4.4 Crash analysis: (a) mesh at t=0 ms; (b) mesh at t=20 ms; (c) mesh at t=40 ms.

Earthquake Response of Soil

It is important to involve the soil skeleton or the interaction of the matrix with the water contained in the hole in the earthquake project. Here we need to solve the transient nonlinear equation. Such as mixing problems, variables include displacement and pore pressure in fluid p.

In figure 4.5 we shown an example of comparison of calculation and a centrifuge model presented at a 1993 workshop known as VELACS . This figure shows the displacements of a big retaining wall after the passage of an earthquake, which were measured in the centrifuge and also calculated.

In Figure 4.5, we show a comparative example of a centrifuge model called VELACS at the 1993 workshop. The figure shows the calculated displacement of the large retaining wall after the earthquake measured in the centrifuge.

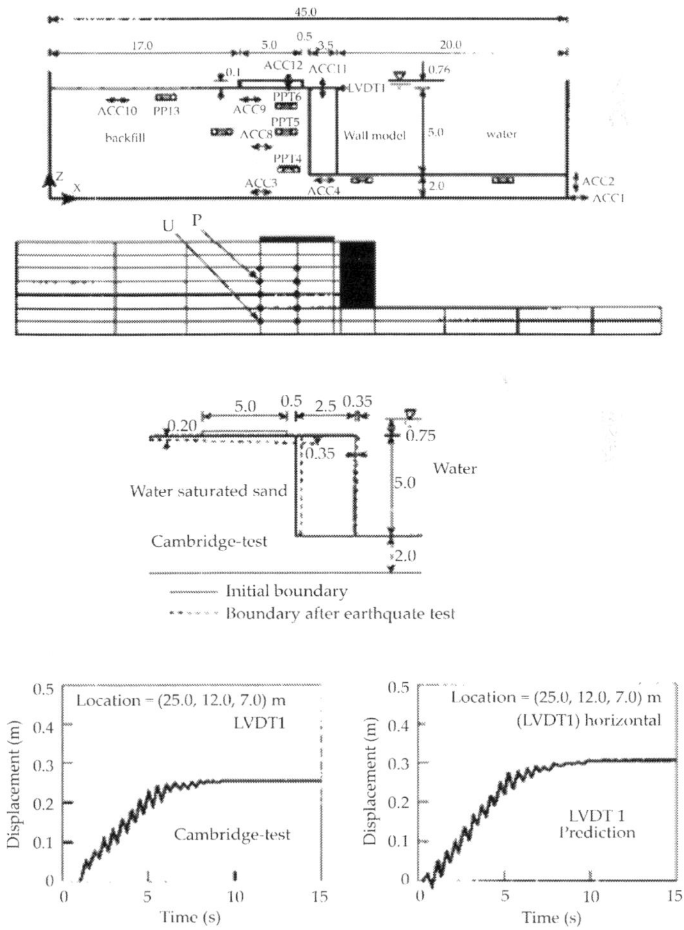

Figure 4.5 Retaining wall subjected to earthquake excitation: comparison of experiment (centrifuge) and calculation.

CHAPTER 5

FINITE ELEMENT METHOD IN STRUCTURE ANALYSIS

1D FINITE ELEMENTS FOR AXIALLY LOADED

Introduction

Some definition of the 1D FEM sample is given in this chapter. Here shows How FEM works on solving this rod problem.

Axially Loaded Rod

As shown in Figure 5.1, given an example of a rod, the rod's length is l, besides a distributed axial load density $t_x(x)$ works on the rod body, a special load F_{x_i} applied on the point. There are p points on the rod, noted as x_i. The displacements of each points written as u_j at x_j. The displacements contribute to an strain $å(x) = \frac{du}{dx}$, the stress in the rod written as ó , the relationship of stress and strain can be described by Hooke Law as following.

$$ó = Eå = E\frac{du}{dx} \quad (5.1)$$

E - tYoung's modulus.

As shown from Figure 5.1, the force $\varkappa$ is getting from the stress and the cross section areas. The value can be calculated from integration.

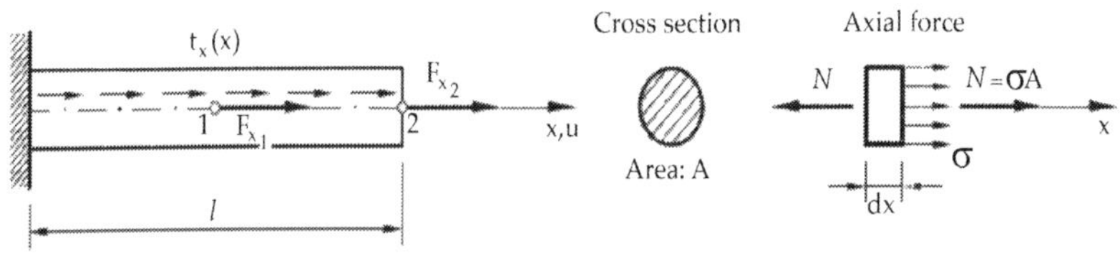

Figure 5.1: Axially loaded rod.

For homogeneous material

$$\aleph = \iint_A \sigma dA = A\sigma = EA\frac{du}{dx} \quad (5.2)$$

If the material's shape is equilibrium, PVW should be satisfied, PVW means Principle of Virtual Work. It constrains the stress and forces. It can be calculated by the follow equation.

$$\iiint_V \delta\varepsilon\sigma dV = \iint_0^l \delta u t_x dx + \sum_{i=1}^{p} \delta u_i F_{x_i} \quad (5.3)$$

δ_u -Virtual displacement of one point

δ_ε -Virtual strain of one point

δ_{u_i} -Virtual displacement of load acting point

t_x -Distributed axial force

V -Volume of the rod

The left part of the equation is the internal and external virtual work caused by the real stress, while on the right part of the equation is the one caused by the external loads.

Another form of equation (5.3) is the shown below by the integarion of the areas.

$$\iiint_V \delta\varepsilon\aleph dV = \iint_0^l \delta u t_x dx + \sum_{i=1}^{p} \delta u_i F_{x_i} \qquad (5.4)$$

$\aleph$ -Axial force leads to displacement

So the major problem for the rod is to get the displacement solution, meanwhile meet the need of equation (5.4) and B.C. In another word, find a solution of sdisplacement, which is quite near the exact solution of displacement and meet the requirement of equation (5.4) and boundary conditions.

During this approximating process to get the displacement, the easiest way is to use polynomials to define the needed elements. So, firstly, discretizing the rod to meshes in finite numbers. The meshing could be written as,

$$\begin{aligned} u(x) \cong \hat{u}(x) &= a_0 + a_1 x + a_2 x^2 + \cdots + a_n x^n \\ &= \sum_{i=1}^{n} a_i x^i \end{aligned} \qquad (5.5)$$

In the above equation:

n-Number of nodes , the displacement of the nodes is unknown.

$a_0, a_1 \ldots, a_n$ -Constants, depend on the displacements.

The top sign of the solution will be removed in the next, so the Equation (5.5) could be written by the follow form

$$u(x) = N_1^{(e)}(x)u_1^{(e)} + N_2^{(e)}(x)u_2^{(e)} + \cdots + N_n^{(e)}(x)u_n^{(e)} = \sum_{i=1}^{n} N_i^{(e)}(x)u_i^{(e)} \quad (5.6)$$

$N_1^{(e)}(x), N_2^{(e)}(x), \ldots, N_n^{(e)}(x)$ -polynomial functions which interpolating in the domain of each element.

e -each element;

$u_i^{(e)}$ -displacment value of the node

$N_i^{(e)}(x)$ is also called shape function, it will interpolate each element the displacement. In the definition of shape function, all the other value of nodes is equal zero except the node I, $u(x_i) = u_i^{(e)}$.

It can replace each element's displacement in PVW. The node displacement of the finite element can be used to present the equilibrium equations. This algebra formulation could be described by the standard matrix form.

$$K\,a = f \quad (5.7)$$

In here, in the rod system or the similar system, K is called the stiffness matrix o the finite elements, a is the displacement of the nodes, f is the vector of the equilibrium force of each node. By the combination of each elements' stiffness matrix, the K and f could be obtained.

If it is in the rod structured system, solving the equation (5.7) can produce all the displacement value of every node. The axial stress, axial force, axial norm force could also be noticed during this process. These concepts will be used to analysis the rod problem with constant cross section areas and end point force, by using two systems with two linear rod componente.

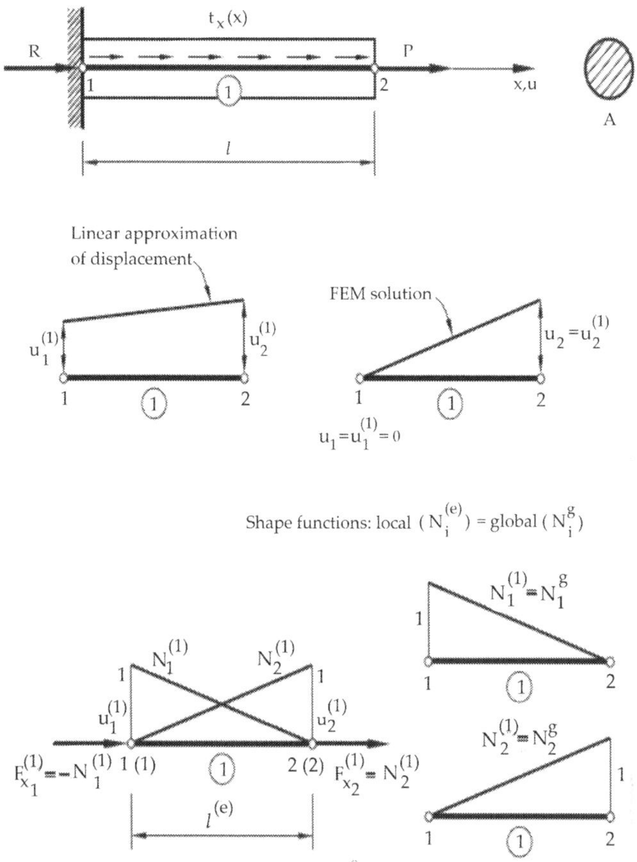

Figure 5.2: Axially loaded rod. Discretization in a single 2-noded element. Number in brackets at node denotes global node number

Constant Cross Section, Discretization in One Linear Rod Element

We can consider a situation shown as Figure 5.2, a rod with a distributed axial force *tx*(*x*) and an external point force P. The cross section is constant circular. The displacement can be defined as the following equations. The linear rod is divided into finite elements for each element there is two nodes.

$$u(x) = \alpha_0 + \alpha_1 x \tag{5.8}$$

The value of u(x) is getting from $u_1^{(1)}$ and $u_2^{(1)}$, $u_1^{(1)}$ and $u_2^{(1)}$ here represent node 1 and 2 respectively

$$\mathrm{u}\left(x_1^{(1)}\right) = u_1^{(1)} \text{ and } \mathrm{u}\left(x_2^{(1)}\right) = u_2^{(1)} \tag{5.9}$$

Node 1 and node 2 have the coordinates, written as $x_1^{(1)}$ and $x_2^{(1)}$.

The top numbers in equation. (5.9) means the element number.

The following equations are getting from the above two equations-Eqs. (5.8) and (5.7)

$$\begin{aligned} u_1^{(1)} &= \alpha_0 + \alpha_1 x_1^{(1)} \\ u_2^{(1)} &= \alpha_0 + \alpha_1 x_2^{(1)} \end{aligned} \tag{5.10}$$

α_0 and α_1 are defined by:

$$\begin{aligned} \alpha_0 &= \frac{u_1^{(1)} - u_2^{(1)}}{x_1^{(1)} - x_2^{(1)}} \\ \alpha_1 &= \frac{x_2^{(1)} u_1^{(1)} - x_1^{(1)} u_2^{(1)}}{x_1^{(1)} x_2^{(1)}} \end{aligned} \tag{5.11}$$

Substituting equation(5.11) into (5.8), equation(5.8) could be written as the following one:

$$\mathrm{u} = N_1^{(1)}(x) u_1^{(1)} + N_2^{(1)} u_2^{(1)}$$

Node 1 and 2's shape function is represented by $N_1^{(1)}$ and $N_2^{(1)}$, the value of them is given by the following equations.

$$N_1^{(1)}(x) = \frac{x_2^{(1)} - x}{l^{(1)}}$$

$$N_2^{(1)}(x) = \frac{x - x_1^{(1)}}{l^{(1)}}$$

(5.12)

Here $l^{(1)} = x_2^{(1)} - x_1^1$ shows how long the element is. It is obtained from Equation (5.12). While shape functions $N_i^{(1)}$ $(i = 1;\ 2)$ change linearly inside the element, taking the value to be one at node i, in other node it is zero. This is a nature result of the shape functions' interpolator property. It can guarantee one condition which is the displacement $u(x)$ set in $u^{(1)}_1$ and $u^{(1)}_2$ at the element nodes. It can be applied in most cases of the shape function geometry, as it will be frequently seen in many finite element method books.

It is important to know the differences between how to number local and global indexes. So let us make it clear first then we can do further study. In Table 5.1 there is an example of how to number the indexes for the nodes, the example's nodal coordinate as well as the nodal displacement is presented in figure 5.2.

It should be noticed that for this case there is only one element, so the local and global indexes are same.

The derivatives of each shape functions are shown in the following equations

$$\frac{dN_1^{(1)}}{dx} = -\frac{1}{l^{(1)}}$$

$$\frac{dN_2^{(1)}}{dx} = -\frac{1}{l^{(1)}}$$

(5.13)

Table 5.1: Local and global parameters for the example 1

Element	node		coordinate		displacement	
	local	global	local	global	local	global
1	1	1	$x_1^{(1)}$	x_1	$u_1^{(1)}$	u_1
	2	2	$x_2^{(1)}$	x_2	$u_2^{(1)}$	u_2

Each nodes inside one element has their axial strain and force, these value could be getting by the following definition.

$$\varepsilon^{(1)} = \left(\frac{du}{dx}\right)^{(1)} = \frac{dN_1^{(1)}}{dx}u_1^{(1)} + \frac{dN_2^{(1)}}{dx}u_2^{(1)}$$
$$= -\frac{1}{l^{(1)}}u_1^{(1)} + \frac{1}{l^{(1)}}u_2^{(1)} \qquad (5.14\_1)$$

$$\aleph^{(1)} = (EA)^{(1)}\varepsilon^{(1)} = (EA)^{(1)}\varepsilon^{(1)}$$
$$= (EA)^{(1)}\left[\frac{dN_1^{(1)}}{dx}u_1^{(1)} + \frac{dN_2^{(1)}}{dx}u_2^{(1)}\right]$$
$$= (EA)^{(1)}\left[-\frac{1}{l^{(1)}}u_1^{(1)}\right.$$
$$\left.+\frac{1}{l^{(1)}}u_2^1\right] \qquad (5.14\_2)$$

It is easy to see, this linear approximation for the displacement need to obey the rule for the axial strain and the axial force on the element.

Since we could obtain the equilibrium equations' derivation. Consider one situation, we set all the elastic modulus, cross section area, distributed loading to be constant, then the equations shown bellowing can be got.

$$K^{(1)} = \left(\frac{EA}{l}\right)^{(1)}\begin{bmatrix} 1 & -1 \\ -1 & 1 \end{bmatrix}$$
$$f^{(1)} = \begin{Bmatrix} f_{x_1}^{(1)} \\ f_{x_2}^{(1)} \end{Bmatrix} = \frac{(lt_x)^{(1)}}{2}\begin{Bmatrix} 1 \\ 1 \end{Bmatrix} \qquad (15)$$

In the content we showed before, bar structured situation has been illustrated. So the global governor equations $\mathrm{Ka} = \mathrm{f}$ can also be got by the similar process using nodal load balance. Every node in the discrete mesh there is:

$$\sum_{e} F_{x_i}^{(e)} = P_{x_j} \qquad ,j = 1, N$$

(5.16)

If the elements have some global node, these elements plus together.

$F_{x_i}^{(e)}$ -Nodal force, which is the sum of each element

P_{x_j} -external load.

If the structure only have one element, Equation (5.16) is given. (also shown in figure 5.2)

$$node1: F_{x_1}^{(1)} = R$$

$$node2: F_{x_2}^{(1)} = P$$

If replacing the values of the nodal forces, by applying the Table 5.1, the follow global equilibrium equations can be obtained.

$$\left(\frac{EA}{l}\right)\begin{bmatrix} 1 & -1 \\ -1 & 1 \end{bmatrix}\begin{Bmatrix} u_1 \\ u_2 \end{Bmatrix} = \begin{Bmatrix} R + \frac{lt_x}{2} \\ P + \frac{lt_x}{2} \end{Bmatrix}$$

Or

$$\mathrm{Ka} = \mathrm{f} \qquad (5.17)$$

K-Global stiffness matrix

a- Displacement matrix, while it is only a vector with all the nodal displacement.

f-Force matrix, It is also a vector. External loads working on a node, which are assigned directly to the responding location.

Eq.(5.17) can be solved by applying the boundary condition $u_1 = 0$, here,

$$u_2 = \frac{l}{EA}\left(P + \frac{lt_x}{2}\right) \quad ; \quad R = -\left(P + lt_x\right) \tag{5.18}$$

The reaction R in equation (5.18) can be gotten from the first row of equation (5.17).

In the most case, the reaction at the described nodes can be calculated as

$$\mathrm{r} = \mathrm{Ka} - f^{ext} \tag{5.19}$$

r-Node reaction

f^{ext} -the assembling of force vector

Stiffness matrix K can be calculated by assembly of the each element $K^{(e)}a^{(e)}$.

Another useful expression for calculating this nodal reaction vector r is shown bellow

$$\mathrm{r} = f_{int} - f^{ext} \tag{5.20}$$

f_{int} - the vector of internal nodal forces that can be gotten by assembling the contributions of the individual elements.

The individual element's nodal force is shown bellow

$$f_{int}^{(e)} = \int_{l^{(e)}}^{\square} \left[\frac{dN_1^{(e)}}{dx}, \frac{dN_2^{(E)}}{dx}\right]^T \aleph^{(e)}\, dx \tag{5.21}$$

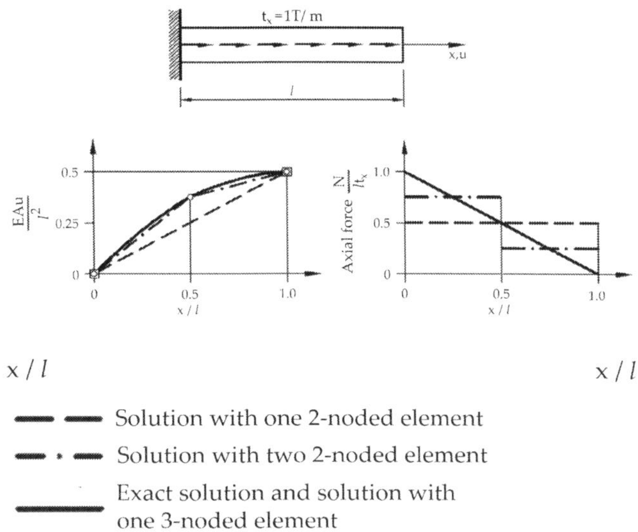

Figure.5.3: Rod under uniformly distributed axial loading. Exact and approximate solutions using one and two linear rod elements and one 3-noded rod element.

The finite element (FE) and the exact results are compared in Figure 5.3 for $P = 0$ and $t_x = 1$, $T = m$.

Be aware of the value of the end displacement u_2 is the exact solution. This is not an always happened result, it only occurs on very few situation.

Inside the rod, one element approximation yields a linear displacement field which is very different from the exact quadratic solution. Also note that the constant axial stress value obtained differs substantially from the linear exact result.

Derivation of the Discretized Equations from the Global Displacement Interpolation Field

A usual description for the displacement interpolation field for the global mesh can be got by overlap of the local approximations for all elements. So we can give a definition of global shape functions which naturally same with the local definition inside each element as shown before. The application of global shape functions result to same results as with the easier local functions. It is significant however to understand the meaningful differences between applying local or global shape functions. For this objective here repeating the one 1-d element problem applying a whole interpolation in the displacement situation.

The 1-d displacement can be described by the one element grid as

$$u(x) = N_1^g(x)u_1 + N_2^g(x)u_2 \tag{5.22}$$

$N_1^g(x)$ and $N_2^g(x)$ are the *global shape functions* of nodes 1 and 2, respectively; u_1 and u_2 are then nodes 1and nodes 2's displacements.

It should be noticed that the top index *e* for the global displacements has been omitted. Same as equation (2.29), for one node's global function at its node location it takes as one at that node, while it takes *zero* at other positions. It can deduce the connection of global and local shape functions shown as following

$$N_i^g = \begin{matrix} N_i^{(e)}(x) & if\ x\ belongs\ to\ element\ e \\ 0 & if\ x\ does\ not\ belong\ to\ element\ e \end{matrix} \tag{5.23}$$

If taking one element case as an example, the global and local shape functions is same (Figure 5.2). so,

$$N_1^g(x) = N_1^{(1)}(x)\ and\ N_2^g(x) = N_2^1(x) \tag{5.24}$$

In the rod shown in figure 5.2, the one dimensional strain can be got by the following equation.

$$\varepsilon = \frac{du}{dx} = \frac{dN_1^g}{dx}u_1 + \frac{dN_2^g}{dx}u_2 \tag{5.25}$$

The fake displacement and the fake rod strain are described as the following equations.

$$\begin{aligned} \delta_u &= N_1^g \delta u_1 + N_2^g \delta u_2 \\ \delta\varepsilon &= \frac{dN_1^g}{dx}\delta u_1 + \frac{dN_2^g}{dx}\delta u_2 \end{aligned} \tag{5.26}$$

The Principle of Virtual Work of the rod can be shown by the following form,

$$\int_0^l \left[\frac{dN_1^g}{dx}\delta u_1 + \frac{dN_2^g}{dx}\delta u_1\right](EA)\left[\frac{dN_1^g}{dx}u_1 + \frac{dN_2^g}{dx}u_1\right]dx$$
$$- \int_0^l [N_1^g \delta u_1 + N_2^g \delta u_2] t_x dx$$
$$= \delta u_1 R + \delta u_2 P \qquad (5.27)$$

As the virtual displacement can be eliminated, equation (5.27) can be deduced to

$$\left(\int_0^l EA\begin{bmatrix} \frac{dN_1^g}{dx}\frac{dN_1^g}{dx} & \frac{dN_1^g}{dx}\frac{dN_2^g}{dx} \\ \frac{dN_2^g}{dx}\frac{dN_1^g}{dx} & \frac{dN_2^g}{dx}\frac{dN_2^g}{dx}\end{bmatrix}dx\right)\begin{Bmatrix} u_1 \\ u_2 \end{Bmatrix} - \int_0^l \begin{Bmatrix} N_1^g \\ N_2^g \end{Bmatrix} t_x \, dx$$
$$= \begin{Bmatrix} R \\ P \end{Bmatrix} \qquad (5.28)$$

The equations shown below are significant for calculation.

$$\left.\begin{aligned} N_1^g &= N_1^{(1)} \\ N_2^g &= N_2^{(1)} \\ \frac{dN_1^g}{dx} &= \frac{dN_1^{(1)}}{dx} \\ \frac{dN_2^g}{dx} &= \frac{dN_2^{(1)}}{dx} \end{aligned}\right\} \quad 0 \le x \le l \qquad (5.29)$$

By equation (5.30) the follow could be got.

$$\int_0^l \frac{dN_1^g}{dx}\frac{dN_1^g}{dx}dx = \int_0^{l^{(1)}} \left(\frac{dN_1^{(1)}}{dx}\right)^2 dx = \frac{1}{l}$$
$$\int_0^l \frac{dN_1^g}{dx}\frac{dN_2^g}{dx}dx = \int_0^{l^{(1)}} \frac{dN_1^{(1)}}{dx}\frac{dN_2^{(1)}}{dx}dx = -\frac{1}{l}$$
$$\int_0^l \frac{dN_2^g}{dx}\frac{dN_2^g}{dx}dx = \int_0^{l^{(1)}} \left(\frac{dN_2^{(1)}}{dx}\right)^2 dx = \frac{1}{l}$$
$$\int_0^l N_i^g dx = \int_0^{l^{(1)}} N_1^{(1)} dx = \int_0^{l^{(1)}} N_2^{(1)} dx = \frac{l}{2} \qquad (5.30)$$

Substituting Eqs.(5.30) into the expression equation(5.28) it will change to *global equilibrium equation.*

Axially Loaded Rod of Constant Cross Section

If it is the same rod shown as before examples, it is discretized into two elements in the rod as shown in Figure 5.4 .The local and global shape functions are also can be seen.

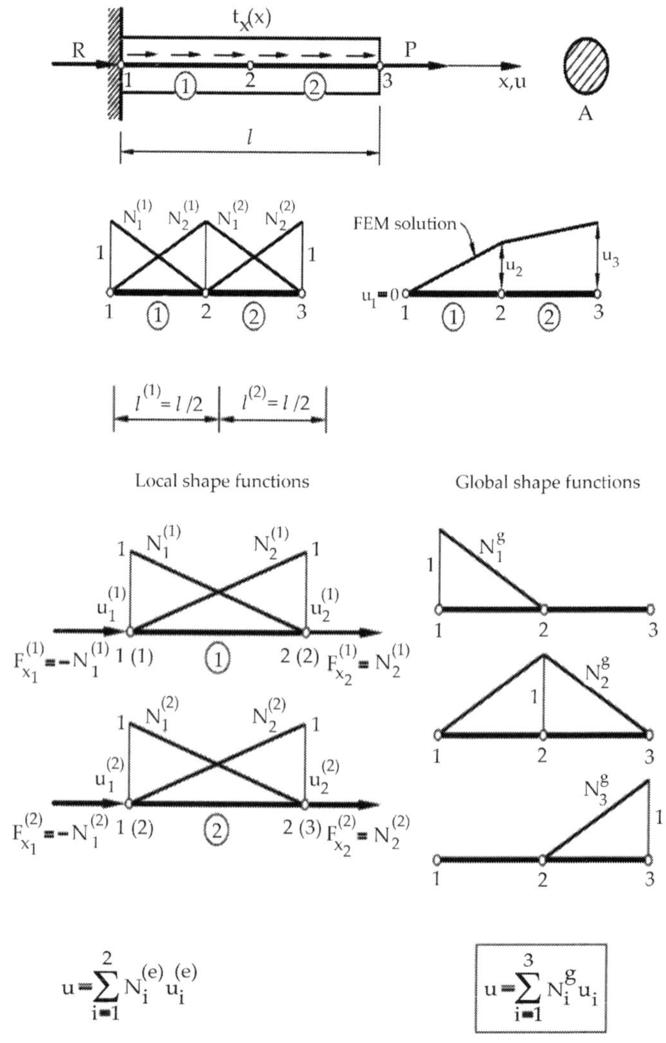

Figure. 5.4: Axially loaded rod. Discretization in two linear elements. Number in brackets at node denotes global node number.

Solution using the element shape functions

The displacements within each element are interpolated as:

$$\textit{Element}\,1 \qquad\qquad\qquad \textit{Element}\,2$$

$$u(x) = N_1^{(1)}(x)u_1^{(1)} + N_2^{(1)}(x)u_2^{(1)} \quad u(x) = N_1^{(2)}(x)u_1^{(2)} + N_2^{(2)}(x)u_2^{(2)} \tag{5.31}$$

The shape functions and their derivatives are

$$N_1^{(1)} = \frac{x_2^{(1)} - x}{l^{(1)}};\ \frac{dN_1^{(1)}}{dx} = -\frac{1}{l^{(1)}}$$

$$N_2^{(1)} = \frac{x - x_2^{(1)}}{l^{(1)}};\ \frac{dN_2^{(1)}}{dx} = \frac{1}{l^{(1)}}$$

$$N_1^{(2)} = \frac{x_2^{(2)} - x}{l^{(2)}};\ \frac{dN_1^{(2)}}{dx} = -\frac{1}{l^{(2)}}$$

$$N_1^{(2)} = \frac{x_2^{(2)} - x}{l^{(2)}};\ \frac{dN_1^{(2)}}{dx} = \frac{1}{l^{(2)}} \tag{5.32}$$

The axial strain in each element is

$$å = \frac{du}{dx} = \frac{dN_1^{(1)}}{dx}u_1^{(1)} + \frac{dN_2^{(1)}}{dx}u_2^{(1)}$$

$$å = \frac{du}{dx} = \frac{dN_1^{(2)}}{dx}u_1^{(2)} + \frac{dN_2^{(2)}}{dx}u_2^{(2)} \tag{5.33}$$

The discretized equilibrium equations are obtained using the PVW as explained in the previous section for the single element case. We find that

$$q^{(1)} = K^{(1)}a^{(1)} - f^{(1)}$$

$$q^{(2)} = K^{(2)}a^{(2)} - f^{(2)} \tag{5.34}$$

The axial strain and the axial force are constant within each element and are obtained as

$$\varepsilon^{(1)} = \left(\frac{du}{dx}\right)^{(1)} = \frac{u_2}{l^{(1)}} = \frac{P + \frac{3lt_x}{4}}{EA}$$

$$\aleph^{(1)} = (EA)^{(1)} \varepsilon^{(1)} = P + \frac{3lt_x}{4}$$

$$\varepsilon^{(2)} = \left(\frac{du}{dx}\right)^{(2)} = \frac{u_3 - u_2}{l^{(2)}} = \frac{P + \frac{lt_x}{4}}{EA}$$

$$\aleph^{(2)} = (EA)^{(2)} \varepsilon^{(2)} = P + \frac{lt_x}{4} \quad (5.35)$$

By the q's components, every element's nodal force could be calculated.

The distribution inside every element of the displacement u and the axial force $u^{(e)}$ has been shown by Figure 2.3 ($P = 0$ and $t_x = 1$T/m). Some approximation improve of the whole displacement region is also can be seen. But, the introduced error in the 1d force is still remarked and its decrease need a better discretization or mesh work. What is more, the nodal smoothing of the constant rod forces over every element. It can be done by simply getting the average of the nodal forces.

Another thing can be seen is that the results obtained for the axial forces or the strains are less precise than those in the displacement distribution. It is a principle rule that is a result from computing the strains or the stresses, they are from the derivatives of the approximate displacement distribution. It increases the result error by the variables.

The similar situation is now be processed by applying the global shape functions.

The rod displacement can be described the two elements mesh together as can be seen from Figure 5.4.

As the virtual displacements can be deleted on both sides, the following matrix system of equations can be written by the below equation.

$$\left(\int_0^l \begin{bmatrix} \frac{dN_1^g}{dx}\frac{dN_1^g}{dx} & \frac{dN_1^g}{dx}\frac{dN_2^g}{dx} & \frac{dN_1^g}{dx}\frac{dN_3^g}{dx} \\ \frac{dN_2^g}{dx}\frac{dN_1^g}{dx} & \frac{dN_2^g}{dx}\frac{dN_2^g}{dx} & \frac{dN_2^g}{dx}\frac{dN_3^g}{dx} \\ \frac{dN_3^g}{dx}\frac{dN_1^g}{dx} & \frac{dN_3^g}{dx}\frac{dN_2^g}{dx} & \frac{dN_3^g}{dx}\frac{dN_3^g}{dx} \end{bmatrix} EAdx\right)\begin{Bmatrix} u_1 \\ u_2 \\ u_3 \end{Bmatrix}$$

$$-\int_0^l \begin{Bmatrix} N_1^g \\ N_2^g \\ N_3^g \end{Bmatrix} t_x dx = \begin{Bmatrix} R \\ 0 \\ P \end{Bmatrix} \tag{5.36}$$

The integrals calculation in equation (5.36) need a connection between the global and local shape functions.

$$\left.\begin{array}{l} N_1^g = N_1^{(1)} \\ \frac{dN_1^g}{dx} = \frac{dN_1^1}{dx} \\ \qquad \frac{dN_2^g}{dx} = \frac{dN_2^g}{dx} \\ \qquad N_3^g = 0 \\ \qquad \frac{dN_3^g}{dx} = 0 \end{array}\right\} \quad 0 \le x \le \frac{l}{2}$$

$$\left.\begin{array}{l} N_1^g = 0 \\ \frac{dN_1^g}{dx} = 0 \\ \qquad \frac{dN_2^g}{dx} = \frac{dN_2^2}{dx} \\ \qquad N_3^g = N_2^2 \\ \qquad \frac{dN_3^g}{dx} = \frac{dN_1^1}{dx} \end{array}\right\} \frac{l}{2} \le x \le l \tag{5.37}$$

This example clearly shows that the application of the global shape functions is less systematic. It also requires more detailed calculation than approach between elements. The differences are more obvious for better meshes. The result of it will be, the combination of the global equations from the elemental description derived by the local shape functions.

Generalization of the Solution with N Linear Rod Elements

The problem solving process explained in the previous sections can easily be summarized that N 2-noded rod elements need to use as discretization for the structure. Every element's stiffness equation could be described as,

$$K^{(e)}a^{(e)} - f^{(e)} = q^{(e)} \qquad (5.38)$$

Solution

The change in cross sectional area is defined by $A = A_0 e^{-\frac{x}{l}}$ where A_0 is the cross sectional area at the clamped end and l is the rod length. The rod is subjected to an axial force acting at the free end.

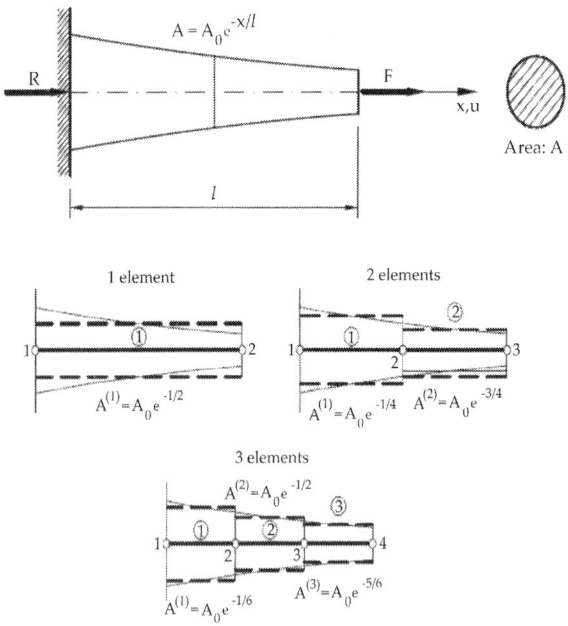

Figure. 5.5: Axially loaded road with exponentially varying cross section. Discretization in 3 meshes of two-noded rod elements.

The exact solution for this simple problem is

$$ó = \frac{F}{A} = \frac{F}{A_0} e^{\frac{x}{l}}$$

$$å = \frac{\sigma}{E} = \frac{F}{EA_0} e^{\frac{x}{l}}$$

$$u(x) = \int_0^x \varepsilon dx = \int_0^x \frac{F}{EA_0} e^{\frac{x}{l}} dx = \frac{Fl}{EA_0}\left(e^{\frac{x}{l}} - 1\right)$$

$$u(1) = \frac{Fl}{EA_0}(e-1) = 1.71828\frac{Fl}{EA_0}$$

$$R = -F$$

The finite element solution has two options: a) get an exact expression using the cross-sectional area, and b) get a constant cross-sectional area within each element. For simplicity, the second option is selected here. You can try the first option as an exercise for this problem.

After the node displacement is obtained, the axial strain or axial stress can be calculated for each element. For example, the center point of element 2 in the three element grid

$$u_4 = 1.71036\frac{Fl}{EA_0}(Error = 0.46\%)$$

$$u_3 = 0.55156u_4 = 0.9432\frac{Fl}{EA_0}$$

$$u_4 = 0.230241u_4 = 0.3938\frac{Fl}{EA_0}$$

$$R = -F$$

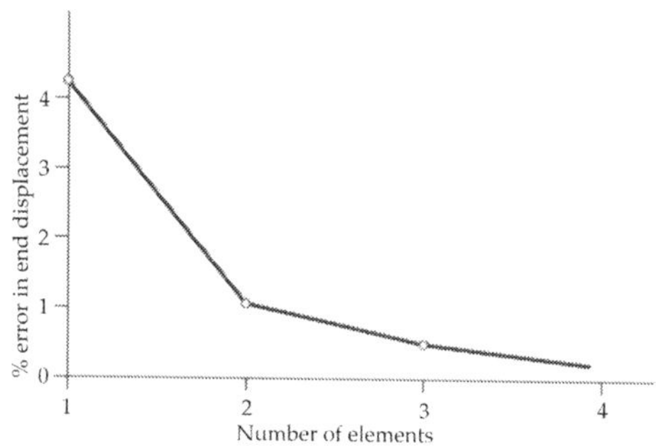

Figure. 5.6: Axially loaded rod with varying cross sectional area. Convergence of the end displacement value and the stress distribution with the number of elements.

The convergence of the end displacement value and the number of elements has already been shown in Figure 5.6. We see that the simple assumption that a constant cross-sectional area results in a smaller percentage error than the two elements is less than 1%.

The displacement and stress distributions along the bars of the three grids are plotted with the exact solution in Figure 5.7. The displacement of nodes in each element, even the linear displacement field, is quite accurate for the three grids. However, the constant axial stress field and the precision exponential solution of each element are slower.

Extrapolation of the Solution with N Linear Rod Elements

Expanding in Taylor series the displacement in the vicinity of a node *i gives*

$$\mathrm{u} = u_i + \left(\frac{\partial u}{\partial x}\right)_i (x - x_i) + \left(\frac{\partial^2 u}{\partial x^2}\right)_i (x - x_i)^2 + \ldots \quad (5.39)$$

If the shape function Ni (x) is a p degree polynomial, it is clear that only the first p term of the Taylor expansion can be precisely approximated to the order p + 1, p + 2, and so on. The approximate error is the order of the first term that is ignored in the above expansion first term disregarded in the above expansion, i.e.

$$\text{error} = u_{exact} - u_{approx} = O(x - x_i)^{p+1} \cong O\left(l^{p+1}\right) \quad (5.40)$$

where $O\left(l^{p+1}\right)$ is read as "of the order of l^{p+1}" and l is the element length.

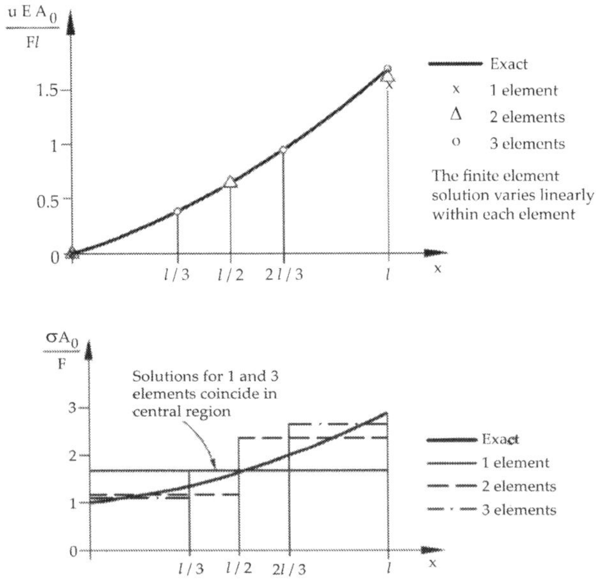

Figure. 5.7: Nodal displacements and axial stress distribution along the rod for the three meshes.

Let us now consider two solutions u^1 and u^2 obtained with two meshes of uniform element sizes l and $l=d$, respectively. The approximate value of u_{exact} as

$$u_{exact} \cong \frac{\left(d^{p+1}\right)u^2 - u^1}{\left(d^{p+1}\right)-1} \tag{5.41}$$

This technique is known as Richardson extrapolation. For the rod of Figure 5.5 we obtain for the end displacement value:

- Extrapolated solution from mesh 1and 2 (d=2);
- Extrapolated solution using meshes 1 and 3 (d=3);
- Extrapolated solution using meshes 2 and 3 (d=1.5)

The Richardson extrapolation method is an effective technique in which the same type of element can be used to improve the dissociation solution obtained from the two grids. This simple process also applies to 2D and 3D problems. Obviously, the enhanced nodal displacement value can be used to obtain the improved solution of the stress field.

Unfortunately, the improvement is independent of node displacement.

Matrix Formulation of the Element Equations

The method described in the previous parts is useful for introducing the basic steps of a finite element method. However, for each node with multiple displacement variables, the matrix formula is more convenient because it allows all variables and algebraic operations to be grouped together in a compact form.

Many expressions used in the future can only be called single elements. For simplicity, the index index "e" node displacement vector representing the element value will be omitted later in addition to several signal element parameters (such as primary geometries (1 (e), A (e), and V (e) (E), the node coordinate vector x (e), the node force vector (f (e); q (e)) and the stiffness matrix K (e). Unlsess otherwise specified, all other parameters appearing in the text Vector and matrix, also known as a single element.

The calculation of the element stiffness matrix K (e) and the equivalent node vector f (e) from the node contributions K (e) and f (e) is simple and economical, and contributes to the organization of the computer program. We will verify this in the whole book. The global stiffness matrix K and the equivalent node vector f are combined from the standard mode element contribution to the global system Ka = f. Once the node displacement a is found, the reaction at the specified node can be calculated.

The definition of internal nodal force vector for the element is shown below

$$f_{int}^{e} = \int_{l^e} B^T \sigma dx \tag{5.42}$$

Steps for the Analysis of a Structure Using the FEM

Step 1. Discretize the structure into a mesh of finite elements.

Step 2. Compute for each element the stiffness matrix and the equivalent nodal force vector due to external loads using expressions of the type K^e, f^e. For two-dimensional and three-dimensional structures, the elemental integrals are caslculated on the element area and volume, respectively.

Step 3. Assemble the stiffness matrix and the equivalent nodal force vector for each element into the global system.

$$\mathrm{Ka} = \mathrm{f}$$

$$\mathrm{K} = A_e K^e$$

$$\mathrm{f} = A_e f^e + p + r(5.43)$$

Here A represents the global compilation of all individual matrices and vectors for each element in the grid. P is the vector of the external point force acting at the node. Once the node displacement is found, r is the vector of the node to be calculated.

The reaction vector r is assembled into f, since the reaction does not affect the solution of the node displacement.

Step 4. The nodal displacements are computed by solving the equation system (2.85a) where the prescribed displacements must be adequately imposed, i.e.

$$\mathrm{a} = K^{-1} f \qquad (5.44)$$

The nodal reactions are obtained at the prescribed nodes.

Step 5. The strains and stresses are computed within each element from the nodal displacements as

$$\begin{aligned} \varepsilon &= \mathrm{Ba} \\ \sigma &= \mathrm{DBa} \end{aligned} \qquad (5.45)$$

The nodal axial forces for each element can be computed from

$$q^e = \left[-\aleph_1^e, \aleph_2^e\right] = K^e a^e - f^e \qquad (5.46)$$

2D SOLIDS, LINEAR TRIANGULAR AND RECTANGULAR ELEMENTS

Introduction

From this part, we will more focus on 2D problem. FEM can be used to the problems which is considered to be ideal two dimensional elastic ones. 2D is study the plane stress or strain instead of linear ones. Most of the definition here is also be used in the 3D problem, which more related with structural. Firstly, let us give an introduction of the how finite element method can be useful for two dimensional or three dimensional problem.

Nearly all structural problem can be solved by finite element method if the two dimensional elasticity assumptions are satisfied. All these problems have one thing common in geometry, it is they are all prismatic shape. According to the prism shape and the loading, the problems can be divided into two major types:

Firstly type is focus on the stress. For the target structure, if one dimension of the three ones is much smaller than the other two. What is more, if the loading is focused in the center part of the plane. This kind of structure has a region of interest in the center part, as shown from Figure 5.8. There are many examples for such type, which more focus on the stress analysis, such as beams, walls which stands the loads which mentioned before, and dams.

The other type is more focus on the strain. For such type of problem, the geometry also has such characteristics, one of the dimension of the three is much larger than the other two. What is more the loads are distributed on the axial direction rather than on the plane, as shown in Figure 5.9. So the region of interest which should be analyzed in the rod's cross section. There are also many examples about this type of problem, like the pipes, frame.

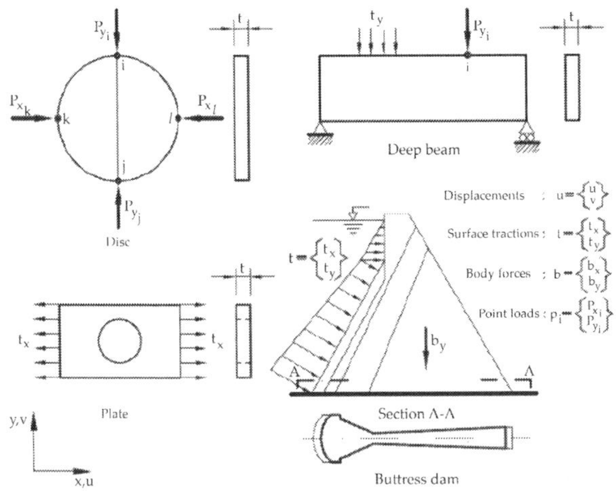

Figure. 5.8: Examples of plane stress problems. Displacement field and loads acting on the middle plane section.

There is a mathematic model for two dimensional problem, so two dimensional solid model could be treated as a three dimensional structural projection. Finite element method could give us an approximation solution to be close the exact one, by using two dimensional solid elements to solve two dimensional problem. The error it introduced is related with the element type and mesh method.

The two dimensional principle can make the finite element analysis be consistent for all type of plane stress and strain possible. The two problem we mentioned above have totally different geometry, we can analysis them respectively.

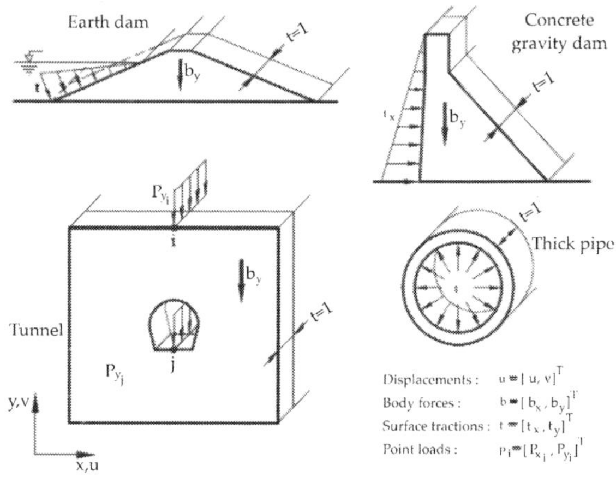

Figure. 5.9: Examples of plane strain problems. Displacement ¯eld and loads acting on a transverse secti

Two Dimensional Elasticity Theory

The two dimensional elasticity theory has an assumption, it is that the transversal section to the prismatic axis will deform just like each other, so the deformation no matter in stress or strain field, by the prismatic axis which we write as Z is omitted. In the 2D problem we only consider a cross section of prismatic axis, which is the X-Y plane. The plane displacement can be written by u(x,y) and v(x,y) which represent the displacement in x axis and y axis separately. They can be written as the following equation:

$$\mathrm{u}(\mathrm{x},\mathrm{y}) = \begin{Bmatrix} u(x,y) \\ v(x,y) \end{Bmatrix} \tag{5.47}$$

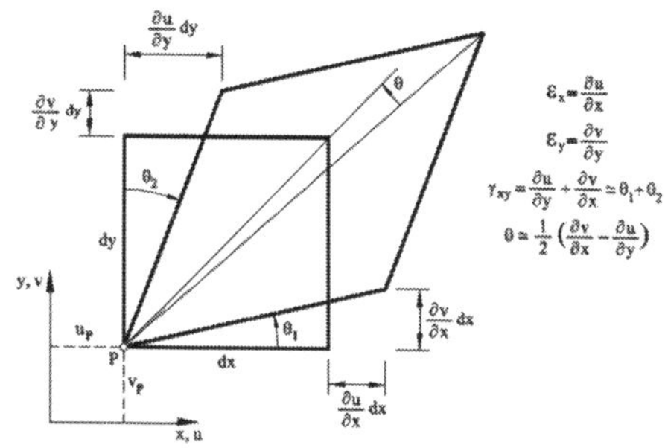

Figure. 5.10: Deformation of an infinitesimal 2D domain and definition of strains

Strain field

By the elastic theory, the following strain description can be written in the displacement field.

$$\varepsilon_x = \frac{\partial u}{\partial x}$$

$$\varepsilon_y = \frac{\partial v}{\partial y}$$

$$\gamma_{xy} = \frac{\partial u}{\partial x} + \frac{\partial v}{\partial y}$$

$$\gamma_{xz} = \gamma_{yz} = 0 \qquad (5.48)$$

The strain in Z axis ε_z is treated as 0 for the 2D strain problem. However the ε_z cannot be treated as 0 for 2D stress problem, although the Z axis stress is considered to be 0. So the Z axis strain need to be settle down before solving the stress or plane situation, however the longitudinal strain (For example, $\sigma_z \varepsilon_z$) can be treated as zero. The strain can be described as the following equation in both situation.

$$å = \left[\varepsilon_x, \varepsilon_y, \gamma_{xy}\right]^T \qquad (5.49)$$

The graphical meaning of the strains for 2D problems is shown in Figure 5.11.

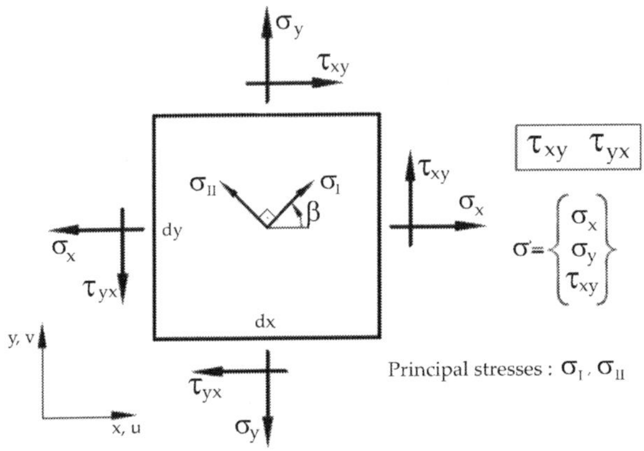

Figure. 5.11: Definition of stresses $\sigma_x, \sigma_y, \tau_{xy}$ and principal stresses σ_I, σ_{II} in 2D solids.

Stress-Strain relationship

The stress strain relation can be obtained from equation (5.49) since τ_{xz} and τ_{yz} are zero. As mentioned in the last part, the stress σ_z is relatively independent with other work, so the stress can described by the following equation which can works for both stress and strain, also shown as Figure 5.11.

$$\sigma = \left[\sigma_x, \sigma_y, \tau_{xy}\right]^T \tag{5.50}$$

What is more, according to the three dimensional elastic theory, if we use the assumption we mentioned before, it is $\sigma_z = 0$ for plane stress, $\varepsilon_z = 0$ for plane strain, and $\gamma_{xz} = \gamma_{yz} = 0$ in both cases. The relationship between stress and strain can also be described by the following equation.

$$\sigma = D\varepsilon \tag{5.51}$$

D is the elastic matrix.

If the sample's material is not isotropic, the temperature effect to the initial strain should not be neglected in the major direction of the material. Then the deformation it introduced to global components ε0 need to be considered. So, if it is like this, the strain τ_{xy}^{0} cannot be treated as zero.

The solid can also be initially subjected to stresses defined by a vector σ^{0}. These initial stresses can have different sources. For instance, if a part of the material is removed from a deformed structure under a set of loads, then automatically a new deformation is originated due to the existence of initial stresses. The sum of all stresses can be got by the sum of the initial stresses. These stresses are from the deformation processes. This can be described by the following equation which is suit for more cases.

$$\sigma = D\left(\varepsilon - \varepsilon^{0}\right) + \sigma^{0} \qquad (5.52)$$

here

$$\sigma^{0} = \left[\sigma_{x}^{0}, \sigma_{y}^{0}, \tau_{xy}^{o}\right]^{T} \qquad (5.53)$$

Where the σ^{0} is the initial stress. Take the tunnel as the example here, for the original stress, the equilibrium of the structure is related with the initial one before starting the cavitation. In manufacture, the importance of initial stress should also be considered and it is always noted as residual stress.

Principle stresses and failure criteria

In stress field, a two dimensional solid can be written by the two major stresses σ_{I} and σ_{II} (shown in Figure 5.10). The principal stresses are the basic components of polynomial, which is unique than others.

$$\det\left(\left[\sigma\right] - \lambda I_{2}\right) = 0 \qquad (5.54\_1)$$

Here

$$\left[\sigma\right] = \begin{bmatrix} \sigma_{x} & \tau_{xy} \\ \tau_{xy} & \sigma_{y} \end{bmatrix} \text{ and } I_{2} = \begin{bmatrix} 1 & 0 \\ 0 & 1 \end{bmatrix} \qquad (5.54\_2)$$

If there is one point fail to satisfy the needs, the value used here is considered to be close to the limit value in the stress field for a two dimensional problem. This can be applied to find the fracture property for the material which is easy to break, for example glasses. Another criteria is try to find the limit for the stress.

Example 5.1: Explain the meaning of the initial strains for the bar in Figure 5.12

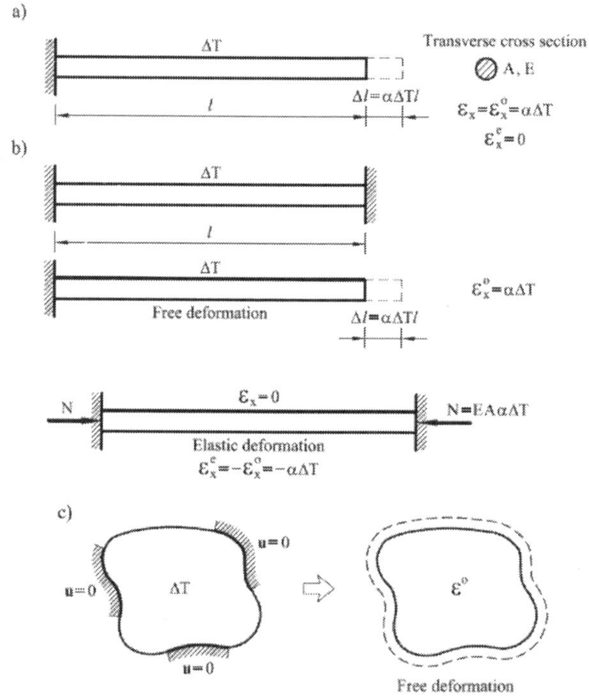

Figure. 5.12: Interpretation of initial thermal strains, a) Clamped/free bar, b) Fully clamped bar, c) 2D solid.

If the rod is fixed at one end point and the other part is in free. Heat the rod in uniform heat, the rod will be longer by uniform increment. The increment of the rod can be written as follows.

$$\Delta l = \alpha \Delta T l$$

and the corresponding "initial" strain is

$$\varepsilon_x^0 = \frac{\Delta l}{l} = \alpha \Delta T$$

Because the rod can be moved, the increased in length is the same value with the heat introduced. The strain from elasticity is zero. As shown from the follow equation.

$$\varepsilon_x^0 = \frac{\Delta l}{l} = \alpha \Delta T$$

The stress from the deformed rod is zero by the above equation.

If the situation is like it shown in Figure 5.12b, firstly, we can assume the rod end can be moved free. How to get the initial strain? The rod will be same with the one mentioned above as shown in Figure 5.12a, in the term of initial deformation. For example, $\varepsilon_x^0 = \alpha \Delta T$. Since the rod is constrained by the two fixed ends, the strain from elasticity can be described by the following equation.

$$\varepsilon_x^e = \varepsilon_x - \varepsilon_x^0 = 0 - \alpha \Delta T = -\alpha \Delta T$$

It can be got that the bar is yield to a uniform axial force $\mathrm{N} = -\alpha \mathrm{EA} \Delta \mathrm{T}$.

Thus, the initial thermal strain can be interpreted as a strain caused by the temperature rise of a point in the free body (Fig. 5.11c). Satisfies the motion (displacement) boundary condition to provide the actual (total) strain value. The difference between the total strain and the initial strain produces a "elastic" strain that is responsible for the actual pressure of the body.

From this example it can be seen that the increase in heat does not move freely in the body's space.

Finite Element Formulation, Three-noded Triangular Element

Firstly, study the simple three-dimensional elements. This is the first element of an analytical structural problem. Prior to the finite element era, Courant successfully solved the differential equations in the two-dimensional domain using the linear polynomial approximation on the triangular region. A few years later, some researchers at Turner et al.

proposed in their classic paper the idea of dividing two dimensional solid domains into simple triangles as a way to analyze solids using matrix structure techniques. This explains why the three-generation triangle is sometimes called the Turner element. This element will soon be welcomed by engineers and is widely used in many structural analyzes of aviation and civil engineering. We note the impact of this element on gravity dams and tunneling studies in practical civil engineering applications. The key to the success of the three generations of triangles lies in its simplicity, allowing assimilation of the finite element analysis and standard matrix methods known to most structural engineers. In contrast, because the linear displacement is similar to the constant strain and stress field, its accuracy is limited. Thus, a fine mesh is required to capture the exact solution in the high gradient region. However, this is not a serious problem because of its general geometry, which is also very suitable for adaptive mesh refinement. In summary, the three-generation triangular element has the ideal feature for applying finite element method to two dimensional solid analysis.

Figure 5.13 shows the cross section of the solids analyzed under plane elasticity assumptions. As usual, the first step is to analyze the domain as a finite element grid discretization. Figure 5.12 shows the selected 3-generation triangular mesh. By using a finer grid, the accuracy of the finite element solution can be significantly improved.

A typical 3-generation triangular element is characterized by the number of nodes of their nodes and their coordinates x; the three nodes have the global number i; the joules; k correspond to the local numbers 1, 2, 3, respectively (Figure 5.13). Use the local number to calculate the correspondence between the element matrix and the vector and the local and global numbers of the assembly process, as in the matrix analysis of the bar structure.

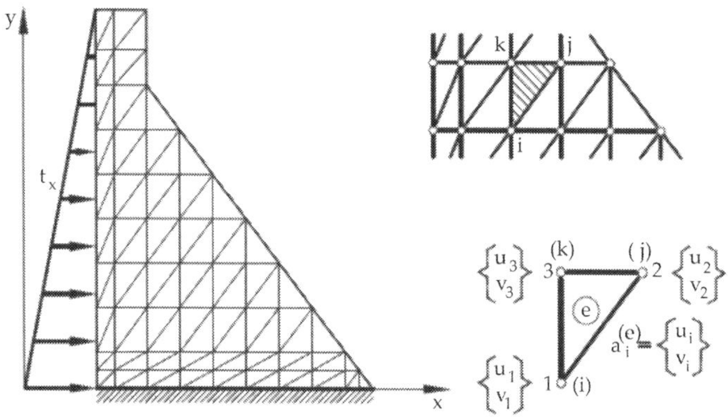

Element	Node		Nodal variables		Nodal coordinates	
	Local	Global	Local	Global	Local	Global
(e)	1	i	u_1 v_1	u_i v_i	x_1 y_1	x_i y_i
	2	j	u_2 v_2	u_j v_j	x_2 y_2	x_j y_j
Area : $A^{(e)}$	3	k	u_3 v_3	u_k v_k	x_3 y_3	x_k y_k

Figure. 5.13: Discretization of a structure in 3-noded triangular elements.

Example 5.2: Analyze the plane structure of the figure 5.13 under self-weight

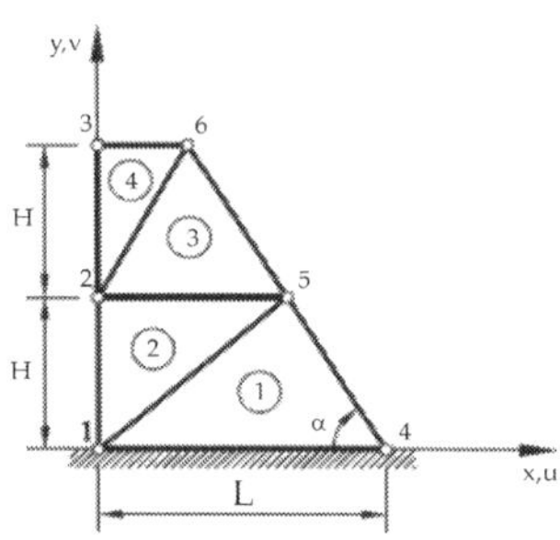

Mesh topology

Element	Nodal connections		
1	1	4	5
2	1	5	2
3	2	5	6
4	2	6	3

Plane strain situation

$u_1 = u_4 = 0$

Figure 5.14: Example 5.2.

The global system of equations has the following form

$$
\begin{array}{c} \\ 1 \\ 2 \\ 3 \\ 4 \\ 5 \\ 6 \end{array}
\begin{array}{c}
\begin{array}{cccccc} 1 & 2 & 3 & 4 & 5 & 6 \end{array} \\
\begin{bmatrix}
(\mathbf{K}_{11}^{(1)}+\mathbf{K}_{11}^{(2)}) & \mathbf{K}_{13}^{(2)} & \mathbf{0} & \mathbf{K}_{12}^{(1)} & (\mathbf{K}_{13}^{(1)}+\mathbf{K}_{12}^{(2)}) & \mathbf{0} \\
 & (\mathbf{K}_{33}^{(2)}+\mathbf{K}_{11}^{(3)}+\mathbf{K}_{11}^{(4)}) & \mathbf{K}_{13}^{(4)} & \mathbf{0} & (\mathbf{K}_{32}^{(2)}+\mathbf{K}_{12}^{(3)}) & (\mathbf{K}_{13}^{(3)}+\mathbf{K}_{12}^{(4)}) \\
 & & \mathbf{K}_{33}^{(4)} & \mathbf{0} & \mathbf{0} & \mathbf{K}_{32}^{(4)} \\
 & & & \mathbf{K}_{22}^{(1)} & \mathbf{K}_{23}^{(1)} & \mathbf{0} \\
 & \text{Symm.} & & & (\mathbf{K}_{33}^{(1)}+\mathbf{K}_{22}^{(2)}+\mathbf{K}_{22}^{(3)}) & \mathbf{K}_{23}^{(3)} \\
 & & & & & (\mathbf{K}_{22}^{(4)}+\mathbf{K}_{33}^{(3)})
\end{bmatrix}
\end{array}
\begin{Bmatrix} \mathbf{a}_1 \\ \mathbf{a}_2 \\ \mathbf{a}_3 \\ \mathbf{a}_4 \\ \mathbf{a}_5 \\ \mathbf{a}_6 \end{Bmatrix}
=
\begin{Bmatrix}
(\mathbf{r}_1+\mathbf{f}_1^{(1)}+\mathbf{f}_1^{(2)}) \\
(\mathbf{f}_3^{(2)}+\mathbf{f}_1^{(3)}++\mathbf{f}_1^{(4)}) \\
\mathbf{f}_3^{(4)} \\
\mathbf{r}_4+\mathbf{f}_2^{(1)} \\
(\mathbf{f}_3^{(1)}+\mathbf{f}_2^{(2)}++\mathbf{f}_2^{(3)}) \\
\mathbf{f}_3^{(3)}+\mathbf{f}_2^{(4)}
\end{Bmatrix}
$$

By eliminating the rows and columns corresponding to the prescribed displacements a1 and a4, the above system can be solved in the usual manner. Once the node displacement is obtained, the corresponding reaction vectors r1 and r4 can be calculated. The constant strain and stress in each element can be found from the known node displacement. It is encouraged that the reader repeat this exercise.

The Four Noded Rectangular Element

Four nodes of rectangles are the simplest quadrilateral elements. This element is developed by Argyris and Kelsey almost simultaneously into 3 nodes of triangles. The general form of quadrilateral thanks tiger. However, the standard 4-node rectangular irregular behavior contributed to a large number of studies, which we will conclude here.

Figure 5.15 shows the use of 4-node rectangular grid discrete deep beam. Let us consider an isolated element with local coordinate system r and s, as shown in Figure 5.15. The four node displacements define the four polynomial interpolation of the displacement field. The simplest interpolation used to satisfy element compatibility and geometric invariance conditions is

$$u(x,y) = \alpha_1 + \alpha_2 r + \alpha_3 s + \alpha_4 r_s$$

$$\tilde{o}(x,y) = \alpha_5 + \alpha_6 r + \alpha_7 s + \alpha_8 r_s \qquad (5.55)$$

Equation (5.55) represents the linear distribution of u and v along each element side, thus ensuring continuity of the displacement field between adjacent elements. Note that the displacement varies within the element with an incomplete quadratic polynomial. The four constants α_i of each displacement component are obtained from the following conditions represented in the local system r and s.

$$\mathrm{u} = u_1 \quad and \ \upsilon = \upsilon_1 \qquad for \qquad r = -a \ , \quad s = -b$$

$$\mathrm{u} = u_2 \quad and \ \upsilon = \upsilon_2 \qquad for \qquad r = -a \ , \quad s = -b$$

$$\mathrm{u} = u_3 \quad and \ \upsilon = \upsilon_3 \qquad for \qquad r = a \ , \quad s = b$$

$$\mathrm{u} = u_4 \quad and \ \upsilon = \upsilon_4 \qquad for \qquad r = a \ , \quad s = b \, (5.56)$$

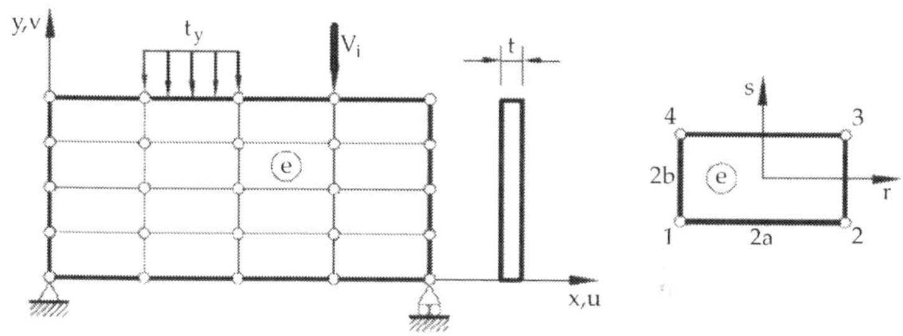

Figure. 5.15: Discretization of a deep beam with 4-noded rectangular elements. Definition of the local axes r and s for an element

Performance of the 3-Noded Triangle and the 4-Noded Rectangle

3 nodes of triangles and 4 nodes of rectangle in pure tension or compression advantage of good performance. In these cases, in general, for the same number of degree of freedoms, the 4-node element is more accurate than the 3 algebraic triangle. However, as mentioned earlier, in the case of bending, the behavior of the two elements is relatively poor. However, the 4-node elements have excellent performance for these problems. In the example shown in Figures 5.16 and 5.17, the thick cantilever under load and the simple support beam analyzed under the weight are clearly visible under the different grids of the 3-node triangle and the 4-node rectangle. All units are in the international system. The comparison

values are: 1) the vertical deviation of the center of the free end of the clamping beam and the lower part of the middle of the simple support shell, 2) the lower ó$_x$ stress component is the two problems. Note that the precision of the 4 generations of rectangles is high.

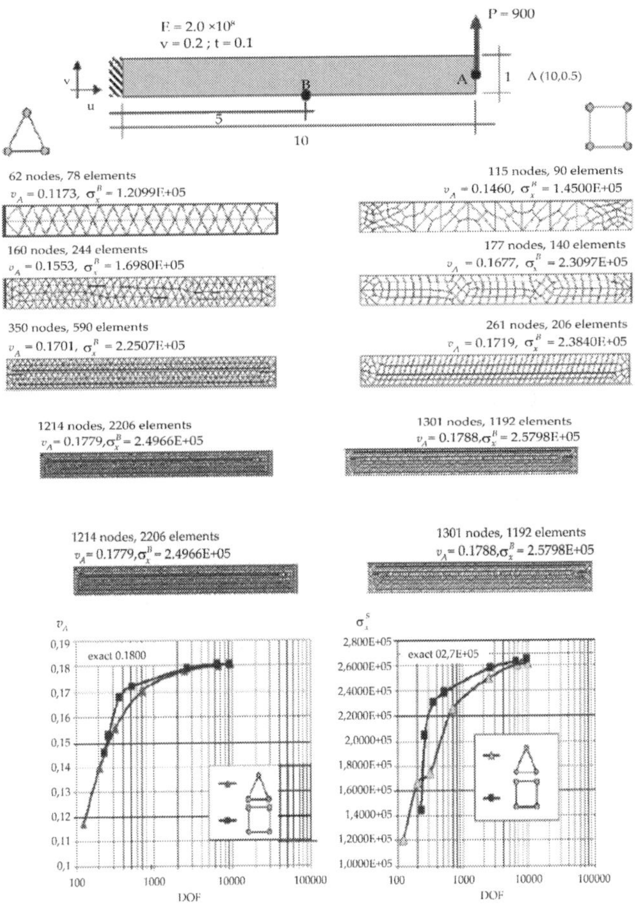

Figure. 5.16: Cantilever beam under end point load. Convergence of the vertical deflection at the free end and the horizontal stress σ_x at the lower fiber of the middle section for unstructured meshes of 3-noded triangles and 4-noded rectangles.

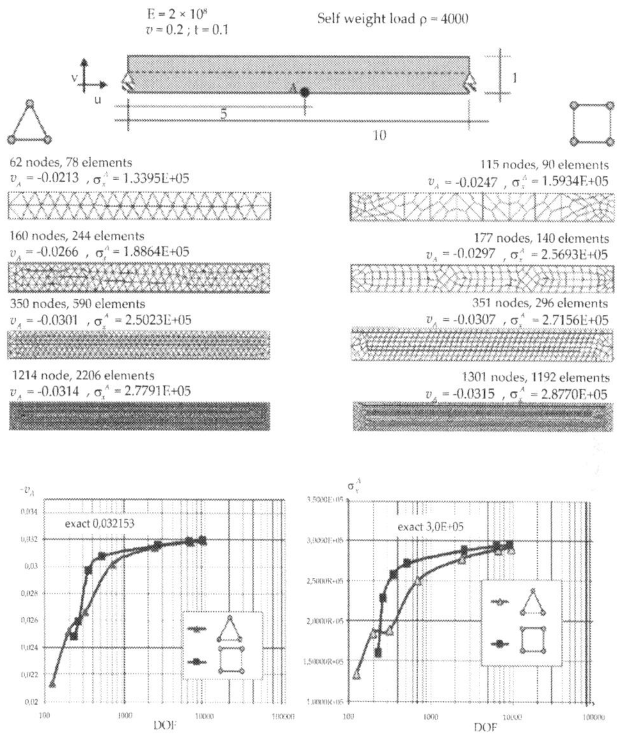

Figure. 5.17: Simply supported beam under self-weight. Convergence of the vertical deflection and the horizontal stress σ_x at the lower fiber of the middle section for different unstructured meshes of 3-noded triangles and 4-noded rectangles.

How precise can the results? The error of two elements can be decreased by applying higher order in displacement. The shortage of the three node mesh is overcame, since it can be used in many situation, especially for different geometries by the triangle elements. It is the most popular method.

3-D SOLIDS

Introduction

By applying the two dimensional stress or strain theory and symmetric model which was introduced in the last section, many structure analysis could be done due to the geometry and load characteristic, this theory also works for the shell and plate models. In this section, we will introduce another general elastic theory which is for the three dimensional problem.

If the solid samples process a heterogeneous material property and random load condition, even the geometry of the sample was not regular, three dimensional analysis need to be operated. As shown in Figure 5.18, there are some classical 3D geometry which requires 3D analysis.

Although 3D analysis seems to be more complex than it in 1D and 2D, it does not bring more basic problem from the concept. The three dimensional analysis is the extension of two dimensional problem. The difference between the two is the three dimensional elements, other analysis work is the same with two dimensional which we already introduced. So this section is more like the repeat of the structural analysis.

Despite the basic theory is not complex, the calculation of three dimensional finite elements was much more than it in two dimensional analysis. It is largely because there is one more dimension than 2d. Besides the calculation time, it need more work to do on the data input and results visualization. As a result, if the problem could be solved by the two dimensional analysis, it is recommended to use this method rather than three dimensional analysis. However, due to many reasons, the three dimensional analysis is irreplaceable in many situations.

The first part of this section shows the basic definitions of the three-dimensional elasticity theory required by applying the finite element method. The details of the element stiffness matrix and the node load vector are given, which is special for simple 4-node tetrahedral elements. And then explain the derivation of the shape function of the hexahedral and tetrahedral elements. Finally, we provided some examples of practical application of 3D finite element analysis.

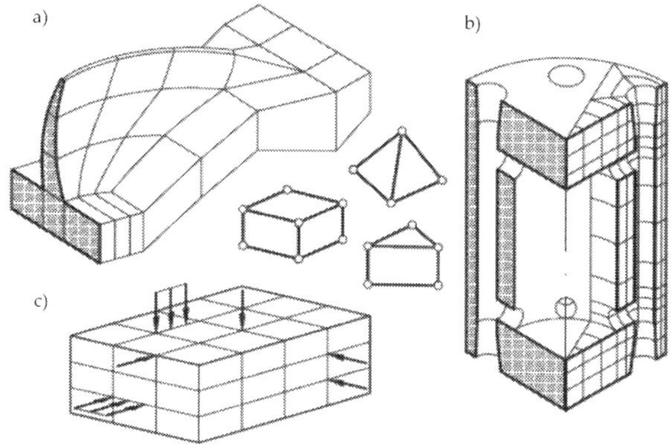

Figure 5.18: Structures which require a 3D analysis: (a) Double arch dam including foundation effects. (b) Pressure vessel. (c) Prismatic solid under arbitrary loading.

Finite Element Formulation, the Four-noded Tetrahedron

The finite element formulation for 3D solid elements will be detailed next. In this part we will give some general expressions to any 3D element with n nodes. Here particularize all matrices and vectors for the simple 4-noded tetrahedron.

The form of $K_{ij}^{(e)}$ for this element is shown in Figure 5.19.

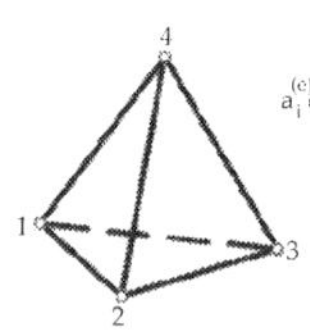

$$a_i^{(e)} = \begin{Bmatrix} u_i \\ v_i \\ w_i \end{Bmatrix} \qquad K^{(e)} = \begin{bmatrix} K_{11}^{(e)} & K_{12}^{(e)} & K_{13}^{(e)} & K_{14}^{(e)} \\ & K_{22}^{(e)} & K_{23}^{(e)} & K_{24}^{(e)} \\ & & K_{33}^{(e)} & K_{34}^{(e)} \\ \text{Symmetrical} & & & K_{44}^{(e)} \end{bmatrix}$$

$$K_{ij}^{(e)} = \frac{1}{36V^{(e)}} \begin{bmatrix} (d_{11}b_ib_j + d_{44}c_ic_j + d_{55}d_id_j) & (d_{12}b_ic_j + d_{44}c_ib_j) & (d_{13}b_id_j + d_{55}d_ib_j) \\ (d_{21}c_ib_j + d_{44}b_ic_j) & (d_{22}c_ic_j + d_{44}b_ib_j + d_{66}d_id_j) & (d_{23}c_id_j + d_{66}d_ic_j) \\ (d_{31}d_ib_j + d_{55}b_id_j) & (d_{32}d_ic_j + d_{66}c_id_j) & (d_{33}d_id_j + d_{55}b_ib_j + d_{66}c_ic_j) \end{bmatrix}$$

d_{ij} are the elements of the constitutive matrix D

b_i, c_i, d_i are the parameters of the shape function N_i

Figure 5.19: Stiffness matrix $K_{ij}^{(e)}$ for the 4-noded linear tetrahedral element.

4-noded tetrahedron

The 4-noded tetrahedron behaves similarly to the 3-noded linear triangle presented in last section. The element has a good ability to model uniform stress fields. However, its accuracy is poor for bending dominated problems, as well as in the presence of high stress gradients and finer meshes are needed in these zones.

We emphasize the intrinsic difficulty of discretizing a solid with tetrahedral. This is a serious problem for the application of the linear tetrahedron to arbitrary shaped solids and particularly when adaptive mesh refinement is used. Much work has been reported in recent years on the development of efficient mesh generators for 3D solids using linear tetrahedral elements.

Other 3D Solid Elements

The formulation presented in Section 8.3 is completely general. This means that the expressions for the stiffness matrix and the equivalent nodal force vector for any 3D solid element coincide with previous equations. The computation of the element integrals simply requires substituting the adequate element shape functions in the expressions for Bi and f(e). In common with 2D solid elements, 3D hexahedra can be of Lagrange or Serendipity types. The shape functions for hexahedral elements are obtained for a right prism in the normalized geometry in terms of the natural coordinates (Figure 5.20) following similar rules as for 2D rec-tangles. The modeling of arbitrary 3D geometries of irregular shape with hexahedra is straightforward using an isoperimetric formulation.

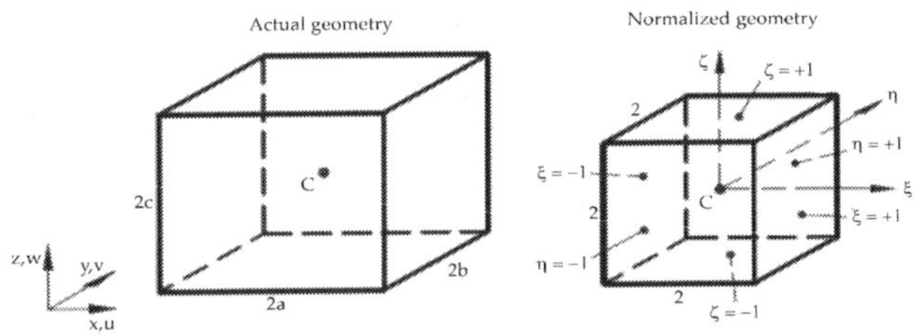

Figure. 5.20: Right prism. Actual and normalized geometries.

The shape functions for Lagrange right prisms are simply obtained by the product of three Lagrange polynomials in 1D. The derivation of the shape functions for 3D Serendipity prismatic elements is more cumbersome, although the same rules given for the 2D case apply.

Like triangles, the shape functions for tetrahedral are complete polynomials and they are more easily expressed in terms of volume and natural coordinates.

Right Prisms

Let us consider a right prism with edges $2a \times 2b \times 2c$, with the natural coordinates ξ, η, ζ defined as shown in figure 5.19. We can write

$$\xi = \frac{(x - x_c)}{a} \quad \eta = \frac{(y - u_c)}{b} \zeta = \frac{(z - z_c)}{c} \tag{5.57}$$

where (x_c, y_c, z_c) are the coordinates of the centroid. Note that the prism becomes a cube in the natural coordinate system (Figure 5.19)

$$\frac{d\xi}{dx} = \frac{1}{a}\frac{d\eta}{dy} = \frac{1}{b}\frac{d\zeta}{dz} = \frac{1}{c} \tag{5.58}$$

and a differential of volume is expressed by

$$dx\,dy\,dz = abcd\xi d\eta d\zeta \tag{5.59}$$

The integration of a function f(x; y; z) over the element is expressed in the natural coordinate system as

$$\iiint_{V(e)} f(x, y, z)dV = \int_{-1}^{+1}\int_{-1}^{+1}\int_{-1}^{+1} f(\xi, \eta, \zeta)abcd\xi d\eta d\zeta \tag{5.59}$$

Since the element is a right prism, the Cartesian derivatives of the shape functions are directly obtained by

$$\frac{\partial N_i}{\partial x} = \frac{1}{a}\frac{\partial N_i}{\partial \xi}$$

$$\frac{\partial N_i}{\partial y} = \frac{1}{b}\frac{\partial N_i}{\partial \eta}$$

$$\frac{\partial N_i}{\partial z} = \frac{1}{c}\frac{\partial N_i}{\partial \zeta} \tag{5.60}$$

The shape functions must satisfy the standard conditions

$$N_i\left(\xi_j,\eta_j,\zeta_j\right) = \begin{cases} 1 & if\, i = j \\ 0 & if\, i \neq j \end{cases} \tag{5.61\_1}$$

and

$$\sum_{i=1}^{n} N_i\left(\xi,\eta,\zeta\right) = 1 \tag{5.61\_2}$$

The shape functions for Lagrange right prisms are obtained by multiplying three 1D Lagrange polynomials as

$$N_i\left(\xi,\eta,\zeta\right) = l_I^i\left(\xi\right) l_1^i\left(\eta\right) l_1^i\left(\zeta\right) \tag{5.62}$$

Where $l_I^i\left(\xi\right)$ is the normalized Lagrange polynomial of Ith degree passing by node i, etc. it is usual to choose the same polynomial approximation in each of the three directions ξ, η and ζ..

The terms contained in the shape functions of prismatic elements are deduced from the Pascal tetrahedron. Figure 5.21 shows the linear and quadratic elements of this family whose shape functions are derived next.

The simplest Lagrange prismatic element is the 8-noded linear prism shown in Figure 5.21.

The nodal shape function is obtained by multiplying the three normalized linear polynomials in ξ, η and ζ., corresponding to the node. In general form

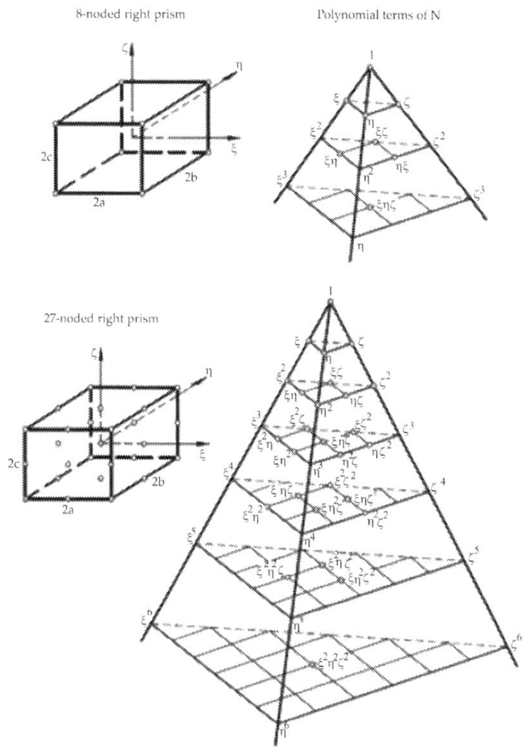

Figure. 5.21: Linear and quadratic right prisms of the Lagrange family. Polynomial terms contained in the shape functions deduced from the Pascal tetrahedron.

Two things need to be noted:

- The shape functions contain a complete linear polynomial in and the incomplete quadratic and cubic termsand (Figure 5.22).

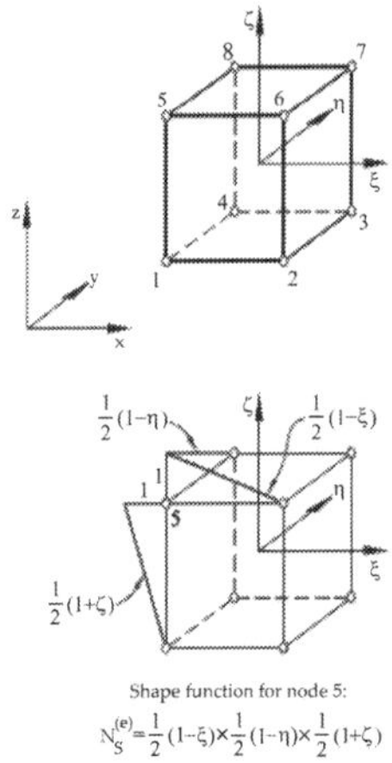

Local node number	Local coordinates ξ_i	η_i	ζ_i
1	−1	−1	−1
2	1	−1	−1
3	1	1	−1
4	−1	1	−1
5	−1	−1	1
6	1	−1	1
7	1	1	1
8	−1	1	1

In general :

$N_i^{(e)} \; \frac{1}{8}(1+\xi_i\xi)(1+\eta_i\eta)(1+\zeta_i\zeta)$

Figure. 5.22: Shape functions for the linear right prism.

- The shape functions satisfy conditions.

The linear hexahedron is a popular element due to the small number of nodal variables, which is attractive for practical 3D analysis.

The linear hexahedron behaves for linear elasticity analysis similarly as the 4-noded rectangle. Its performance is excellent for tension or compression dominated problems, whereas its accuracy is poor for bending dominated situations. This is due to its inability to follow curved deformation patterns and finer meshes are needed to obtain good solutions in these cases.

The behavior of the 8-noded hexahedron can be improved by using reduced integration for the shear stiffness terms, by adding internal or incompatible modes, or by using an assumed strain field, in a similar way to the 4-noded quadrilateral. A popular alternative is to add incompatible internal modes to the original displacement field as

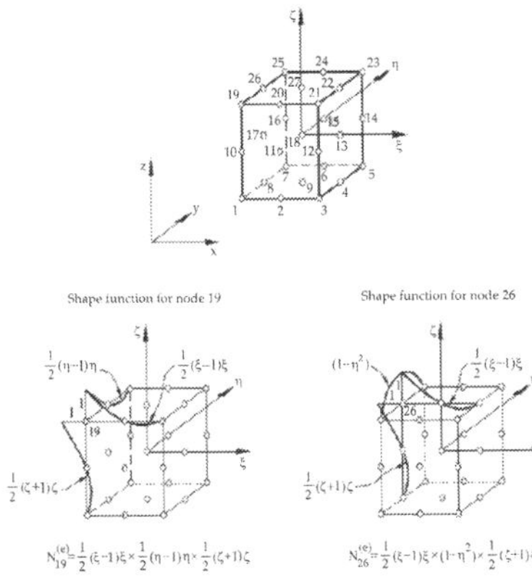

Figure. 5.23: Quadratic right prism of the Lagrange family with 27 nodes. Shape functions for a corner node and a mid-side node.

The enhanced 8-noded hexahedron requires a one point reduced quadrature for the terms corresponding to the incompatible modes, so that the patch test is satisfied. It can be proved that in its right form the enhanced element reproduces pure bending states exactly. The internal incompatible modes can be eliminated by static condensation to yield a stiffness matrix. Hence, the size of the global stiffness matrix does not increase. An explicit form of the element stiffness matrix for right prisms can be obtained as for the 4-noded rectangle.

Other hexahedral elements of the Lagrange family:

The next members of the Lagrange family are the 64-noded cubic prism (4 nodes along each edge) and the 125-noded quartic prism (5 nodes alone each edge). Their shape functions are obtained by the product of three 1D normalized cubic and quartic polynomials respectively. These elements in their right form are in general not competitive versus the analogous Serendipity elements which have less nodal variables.

Serendipity prisms

Serendipity prisms are obtained by extension of the corresponding 2D rectangular Serendipity elements. Figure 5.24 shows the first two elements in the family, i.e. the 8- and 20-noded right prisms.

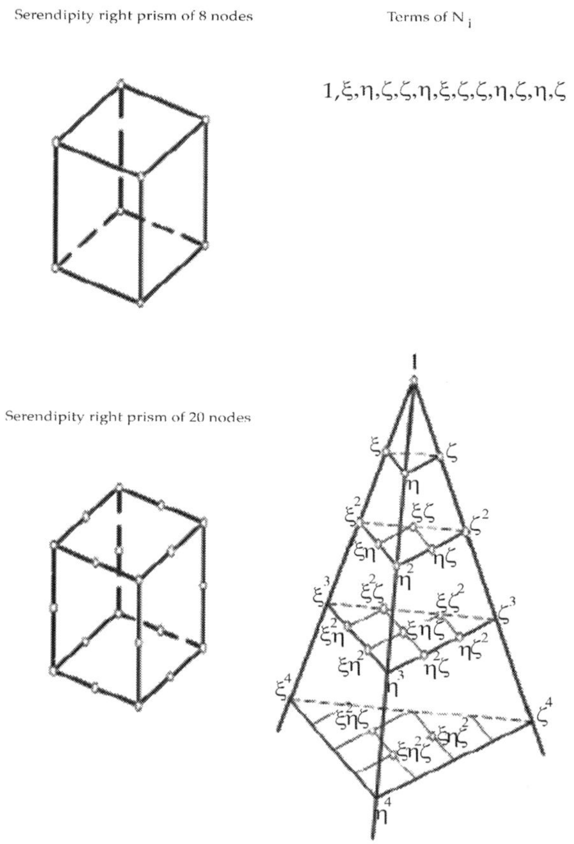

Figure. 5.24: 8-noded and 20-noded Serendipity prisms. Polynomial terms contained in the shape functions deduced from the Pascal's tetrahedron.

20-noded quadratic Serendipity prism

The shape functions are obtained using similar criteria as for the 8-noded rectangle. The shape functions for the side nodes are obtained by multiplying a 1D normalized quadratic Lagrange polynomial and two 1D normalized linear polynomials expressed in the natural coordinates. For the corner nodes, a two-step procedure is followed. The first step involves the derivation of the trilinear function corresponding to the node. This function is subsequently modified so that it takes a zero value at the side nodes. This is achieved by subtracting one half of the values of the shape function of the side nodes adjacent to the relevant corner node under consideration.

Figure 5.23 shows the derivation of the shape function for a side node (20) and a corner node (13).

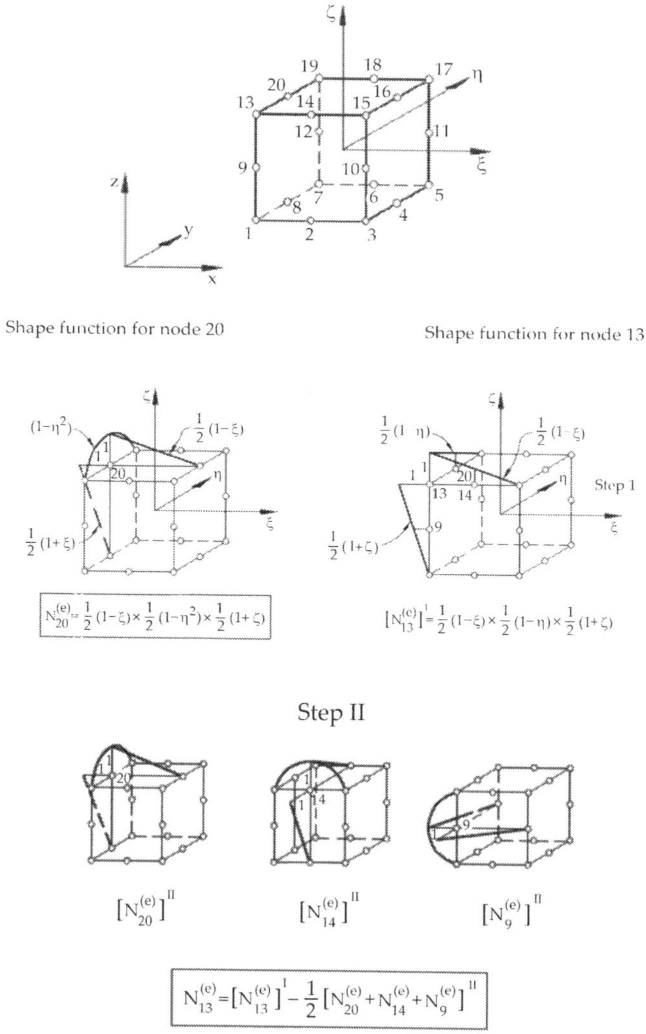

Figure. 5.25: 20-noded quadratic Serendipity prism. Shape functions for a side node and a corner node.

Note that:

1. The shape functions contain a complete quadratic polynomial plus the terms $\xi\eta^2, \xi^2\eta, \xi^2\zeta, \xi\zeta^2, \zeta^2\eta, \eta^2\zeta, \xi\eta\zeta, \xi^2\eta\zeta, \xi\eta\zeta^2$ *and* $\xi\eta\zeta^2$. The shape functions satisfy the conditions.
2. The 20-noded Serendipity prism has the same quadratic approximation as the 27-noded Lagrange prism. This means savings of 21 nodal variables per element, which explains the popularity of the 20-noded right prism 32-noded cubic Serendipity prism:

This element has 8 corner nodes and 24 nodes along the edges as shown in Figure 5.26. The 12 face nodes define a cubic polynomial over each face as for the corresponding quadrilateral element. The shape functions for the side nodes are obtained by multiplying a 1D normalized cubic Lagrange polynomial and two 1D linear polynomials expressed in the natural coordinates. For the corner nodes the starting point is the trilinear shape function, from which a proportion of the shape functions of the side nodes is substracted so that the final shape function takes a zero value at these nodes. The reader is invited to derive the expressions of the shape functions for this element shown in Figure 5.26.

The shape functions contain a complete cubic polynomial (20 terms) plus twelve terms for practical applications. Therefore, the cubic Serendipity prism has the same approximation as the analogous 64-noded Lagrangian prism, but with a substantial reduction in the number of nodal variables. However, this element is less popular than the 20-noded prism.

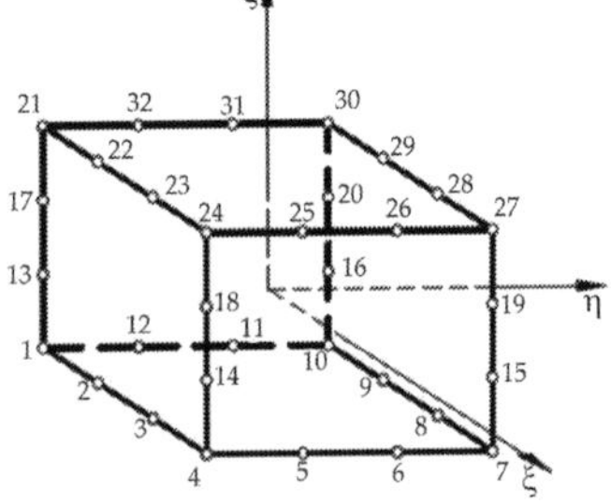

Shape functions

$$N_i = \frac{9}{64}(1+\xi\xi_i)(1+\eta\eta_i)(1+\zeta\zeta_i)(-\frac{19}{9}+\xi^2+\eta^2+\zeta^2) \qquad i = \begin{cases} 1, \ 4, \ 7, 10 \\ 21, 24, 27, 30 \end{cases}$$

$$N_i = \frac{81}{64}(1-\xi^2)(\frac{1}{9}+\xi\xi_i)(1+\eta\eta_i)(1+\xi\xi_i) \qquad i = \begin{cases} 2, \ 3, \ 8, \ 9 \\ 22, 23, 28, 29 \end{cases}$$

$$N_i = \frac{81}{64}(1+\xi\xi_i)(1\ \ \eta^2)(\frac{1}{9}+\eta\eta_i)(1+\zeta\zeta_i) \qquad i = \begin{cases} 5,\ 6, 11, 12 \\ 25, 26, 31, 32 \end{cases}$$

$$N_i = \frac{81}{64}(1+\xi\xi_i)(1+\eta\eta_i)(\frac{1}{9}+\zeta\zeta_i)(1-\zeta^2) \qquad i = \begin{cases} 13, 14, 15, 16 \\ 17, 18, 19, 20 \end{cases}$$

Figure 5.26: Shape functions for the 32-noded cubic Serendipity prism.

Straight-edged Tetrahedra

Straight-edged tetrahedral elements are a direct 3D extension of straight-sided triangles. Their shape functions are also complete polynomials whose terms can easily be deduced from the Pascal tetrahedron as shown in Figure 5.27.

The shape functions for tetrahedral elements can be written in terms of volume coordinates and/or natural coordinates. The volume coordinates are identified by L1; L2; L3 and L4 and have a similar meaning to the area coordinates in triangles. Each coordinate Li is now defined as the ratio between the volumes of the tetrahedron formed by a point inside the element P and the face opposite to node i, and the total volume (Figure 5.28).

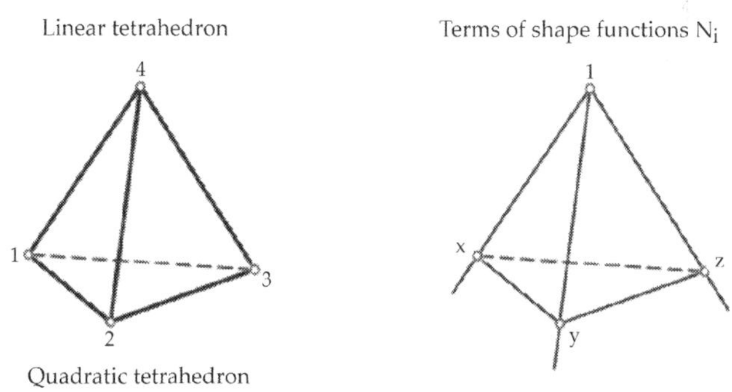

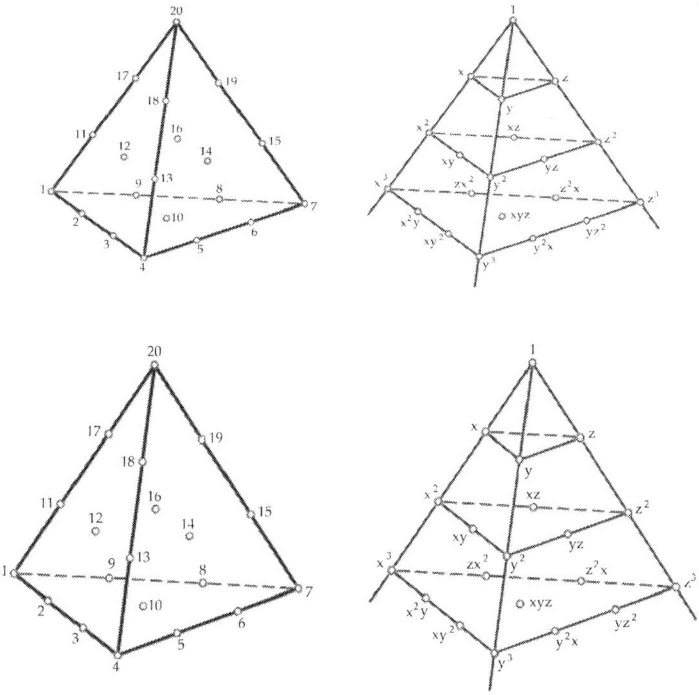

Figure. 5.27: Straight-edged tetrahedral elements: linear (4 nodes), quadratic (10 nodes), cubic (20 nodes). Polynomial terms contained in the shape functions.

The volume coordinates coincide with the shape functions for the 4-noded tetrahedron.

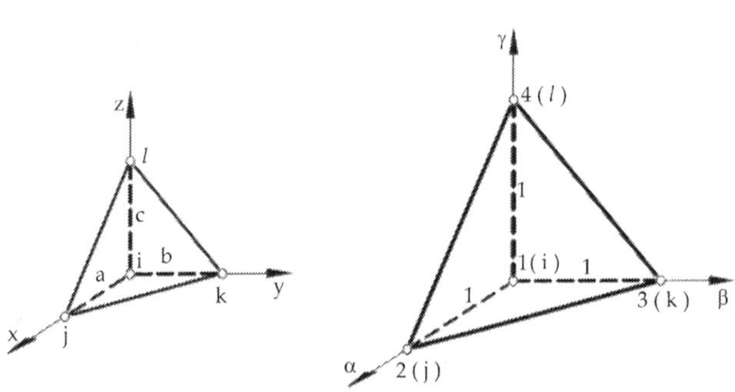

Figure. 5.28: Natural coordinate system in a tetrahedron.

The natural coordinates define a normalized straight tetrahedron of unit right edges and faces for and (Figure 5.26). For a tetrahedron with right edges a, b, c we have.

Shape functions for the 10-noded quadratic tetrahedron.

The nodal values of the generalized coordinates I; J; K; L and of the natural coordinates are shown in figure 5.27.

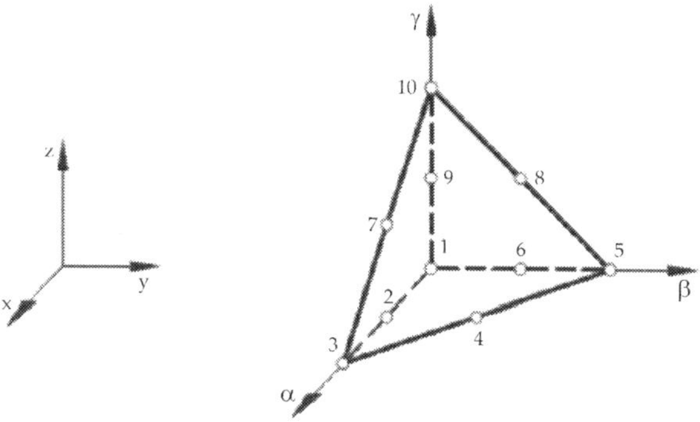

Local node numbers	Coordinates I	J	K	L	Natural coordinates α_i	β_i	γ_i
1	2	0	0	0	0	0	0
2	1	1	0	0	1/2	0	0
3	0	2	0	0	1	0	0
4	0	1	1	0	1/2	1/2	0
5	0	0	2	0	0	1	0
6	1	0	1	0	0	1/2	0
7	0	1	0	1	1/2	0	1/2
8	0	0	1	1	0	1/2	1/2
9	1	0	0	1	0	0	1/2
10	0	0	0	2	0	0	1

Figure. 5.27: 10-noded quadratic tetrahedron. Generalized coordinates I; J; K; L and natural coordinates for each node.

Shape functions for the 20-noded quadratic tetrahedron.

The nodal values of the generalized coordinates I; J; K; L and of the natural coordinates are shown in Figure 5.28.

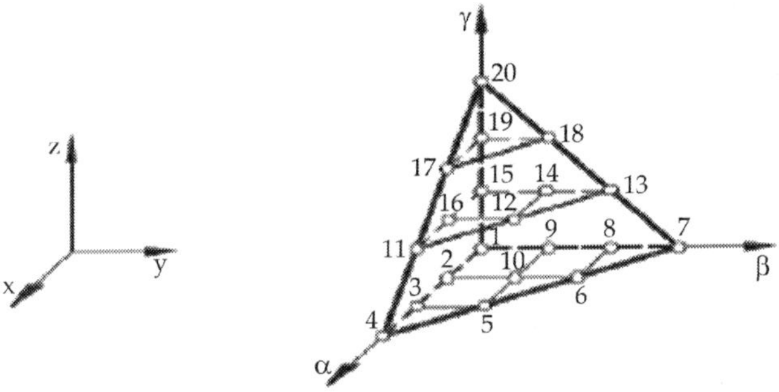

Local node numbers	Coordinates I	J	K	L	Natural Coordinates α_i	β_i	γ_i
1	3	0	0	0	0	0	0
2	2	1	0	0	1/3	0	0
3	1	2	0	0	2/3	0	0
4	0	3	0	0	1	0	0
5	0	2	1	0	2/3	1/3	0
6	0	1	2	0	1/2	2/3	0
7	0	0	3	0	0	1	0
8	1	0	2	0	0	2/3	0
9	2	0	1	0	0	1/3	0
10	1	1	1	0	1/3	1/3	0
11	0	2	0	1	2/3	0	1/3
12	0	1	1	1	1/3	1/3	1/3
13	0	0	2	1	0	2/3	1/3
14	1	0	1	1	0	1/3	1/3
15	2	0	0	1	0	0	1/3
16	1	1	0	1	1/3	0	1/3
17	0	1	0	2	1/3	0	2/3
18	0	0	1	2	0	1/3	2/3
19	1	0	0	2	0	0	2/3
20	0	0	0	3	0	0	1

Figure. 5.28: 20-noded quadratic tetrahedron. Generalized coordinates I; J; K; L and natural coordinates for each node

Computation of Element Integrals

In general, the computation of element integrals is carried out via numerical integration. Useful analytical rules can, however, be derived for straight-edged tetrahedral or right prisms as shown below.

Volume integrals over right prisms can be computed exactly using the local coordinate system xyz. For instance, the integral over a face with z = constant, is first obtained and then a simple line integral in the z direction is performed. The resulting local stiffness matrix K is next transformed to the global axes xyz taking into account the third Cartesian axis. This procedure is not applicable to irregular or curve-sided hexahedra for which the use of an isoparametric formulation and numerical integration is essential.

The integrals over straight-sided tetrahedra have simpler expressions. Thus, the volume integral of a polynomial term expressed in volume coordinates is given by

$$\iiint_{V^e} L_1^k L_2^l L_3^m L_4^n dV = 6V^e \frac{k!\,l!\,m!\,n!}{(k+l+m+n+3)!}$$

(k + l + m + n + 3)!

(5.66)

Similarly, the surface integrals over the element faces can be obtained

The use of natural coordinates does not introduce any additional difficulty. The volume integrals in tetrahedra are computed by

$$\iiint_{V^e} \alpha^k \beta^l \gamma^m dV = 6V^e \frac{k!\,l!\,m!\,n!}{(k+l+m+3)!} \quad (5.67)$$

Curve-sided tetrahedra require an isoparametric formulation and numerical integration.

3-D Isoparametric Elements

Numerical Integration

3D elements with arbitrary geometry can be derived using an isoparametric formulation. The element geometry is defined in the global Cartesian coordinate system in terms of the nodal coordinates, whereas the element integrals are performed in the natural coordinate space over cubes of side length equal to two or straight tetrahedral with right edges of unit length using simple transformations (Figure 5.29).

Let us consider first hexahedral elements. The coordinates of a point within a n-noded element are expressed in isoparametric form as

$$\mathrm{x} = \begin{Bmatrix} x \\ y \\ z \end{Bmatrix} = \sum_{i=1}^{n} N_i \begin{Bmatrix} x_i \\ y_i \\ z_i \end{Bmatrix} = N x^{(e)} \tag{5.68}$$

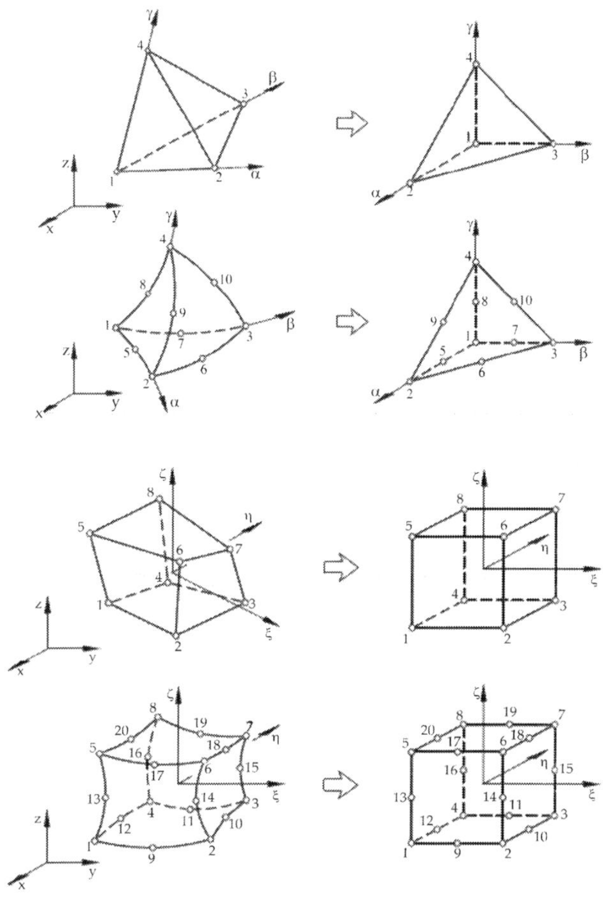

Figure. 5.29: 3D isoparametric elements. Actual and normalized geometries.

Eq. (5.68) relates the cartesian and the natural coordinates. This relationship is uniquely defined if the sign of the determinant of the Jacobian matrix is positive over the element. This is usually fulfilled except for very distorted element shapes.

Examples

Some examples aiming to show the potential of 3D solid elements for solving complex structural problems in engineering are presented. Readers interested in the specific details of each example can contact the authors in the references given.

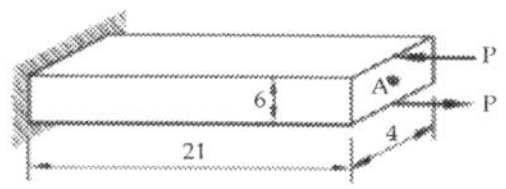

Element	Nodes	Mesh (a×b×c)	Degrees of freedom	Error in vertical displacement of point A
	4	7×3×2×5 = 210 *	84×3 = 252	-25.8%
	8	7×3×2 = 42	84×3 = 252	-10.2%
	10	7×3×2×5 = 210 *	448×3 = 1344	≃1%
	20	7×3×2 = 42	287×3 = 861	Exact (Beam theory)

* Each hexahedron is divided into 5 tetrahedra

Figure. 5.30: Pure bending of a cantilever beam analyzed with different 3D elements.

Example 1

Figure 5.31 shows a perspective view of the geometry of the Mequinenza gravity dam in Spain and the mesh of 20-noded hexahedra used for the analysis. The study of this dam as a 2D solid using 8-noded quadrilaterals. Figure 5.31 displays the contours of the displacement component in the downstream direction and several stress plots for a combination of self-weight and hydrostatic loading. The need for sophisticated graphic visualization techniques for practical 3D analysis such as this one is clear. This topic is treated in Chapter 10 and Annex D.

(a)

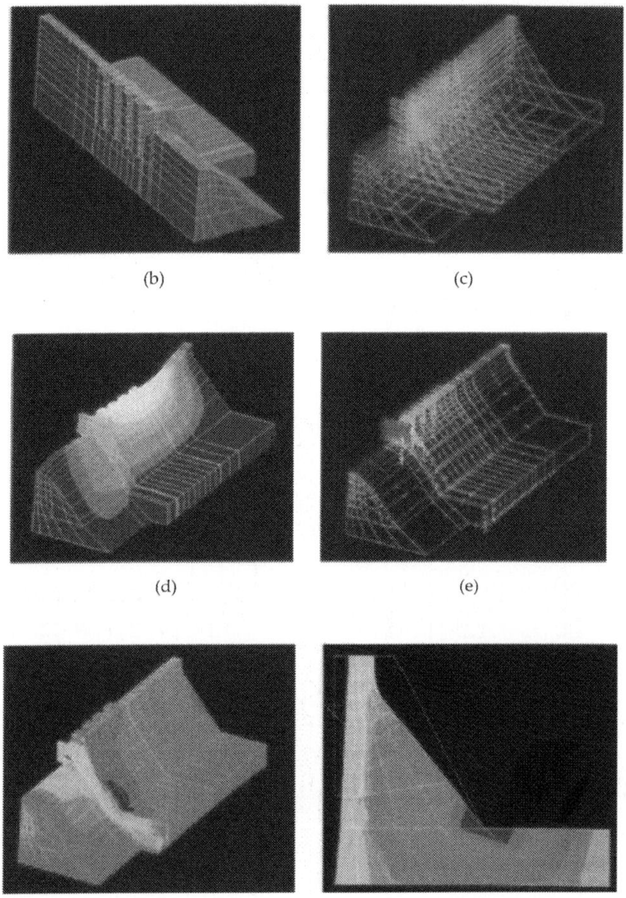

Figure. 5.31: Mequinenza gravity dam analyzed with 20-noded isoparametric hexahedra. (a) Geometry and mesh. (b) Contours of downstream displacement. (c) Principal stresses on dam surface. (d) Contours of transversal stress. (e) Contours of longitudinal stress in a cross section.

Example 2

Figure 5.32a displays the geometry of the Talvachia dam in Italy and the mesh of isoparametric 20-noded hexahedra used for the analysis. The dam foundation has also been discretized. Figure 5.32b shows the contours of the displacement vector modulus for self-weight plus hydrostatic loading. Also the compression principal stresses over the upstream face are dis-played in Figure 5.32c.

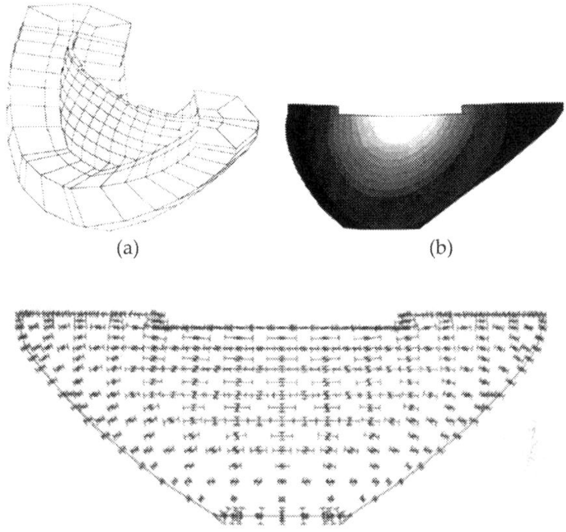

(a) (b)

(c)

Figure. 5.32: Talvachia double curvature arch dam analyzed with 20-noded isopara-metric hexahedra. (a) Geometry and mesh. (b) Contours of displacement modulus and distribution of compression stresses (N/m^2) for self-weight plus hydrostatic loading.

Figure 5.33 shows the analysis of the Melonares arch dam in Sevilla, Spain using again 20-noded Serendipity hexahedra. The interest was the study of the stress field in the spillway zone.

Figure 5.34 shows a snapshot of stress field contours for self-weight load in the surface of an arch dam in Jaen, Spain. The dam and the surrounding terrain were discretized with 4-noded linear tetrahedra in this case.

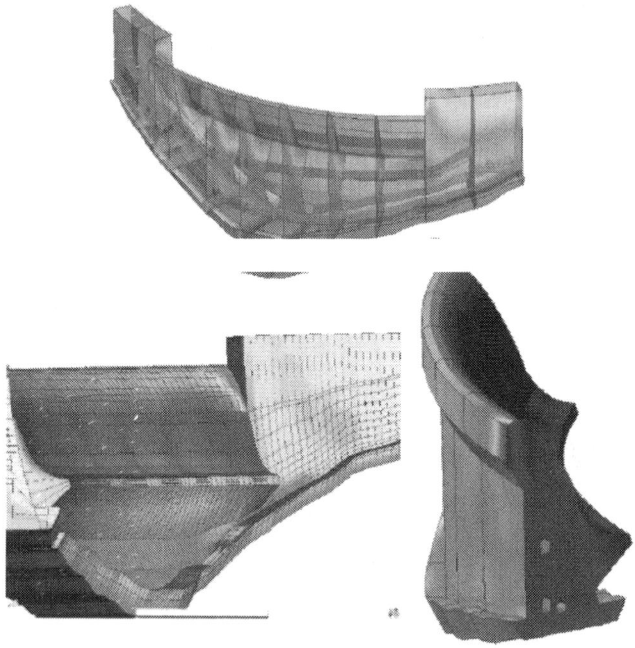

Figure. 5.33: Melonares arch dam. Geometry and discretization with 20-noded Serendipity hexahedra. Detail of stress field in one of the spillways for self-weight load.

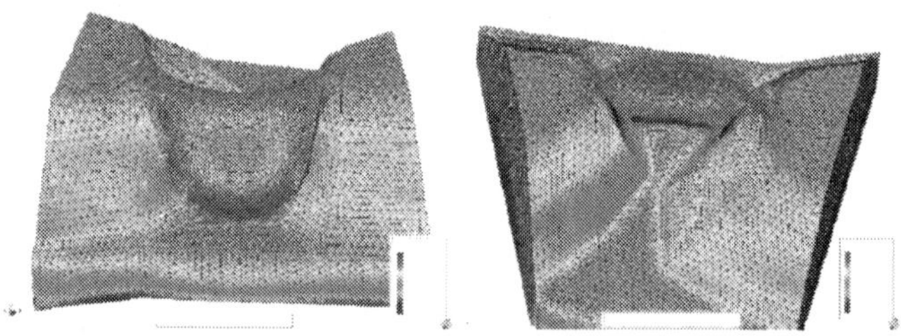

Figure. 5.34: Arch dam in Jaen, Spain. Discretization of dam and terrain with 4-noded tetrahedra. Snapshot of stress field under self-weight loading.

Example 3

Figure 5.35 shows some results of the analysis of a whole at in a building in the city of Barcelona, Spain. Figures 5.35a and 5.35b respectively show the discretization of the walls and upper and lower floors and the rein-forced concrete beams using 20-noded hexahedra. Figures 5.35c show the contours of the displacement modulus plotted on the deformed shape for self-weight load (results have been amplified for visualization purposes).

Reinforced concrete caissons of prismatic cellular shape are commonly used for building piers in harbours. The caisson structure is typically built in the shore and is subsequently launched into the sea and transported as a floating body to the correct position in the pier. The caisson is then sunk into a stable position by filling the internal cells with concrete or rockfill material.

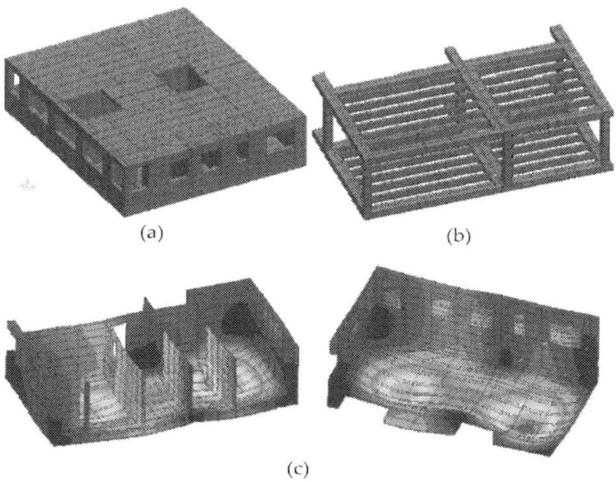

Figure. 5.35: Analysis of a °at in a building. Discretization of the walls and upper and lower floors (a) and the beams (b) using 20-noded hexahedra. Deformed shape and displacement modulus contours for self-weight loading (c). Results have been amplified for visualization purposes.

Figure 5.36a shows the discretization of a cellular caisson built in the Bilbao harbour in Spain (dimensions) using a mesh of 20-noded hexahedra. Half the real geometry of the caisson has been discretized only due to symmetry. Figures 5.36b show results of the de-formed shape of the caisson and the contours of the von Mises stress for a combination of lateral wave loading and internal pressure in some of the cells due to the filling.

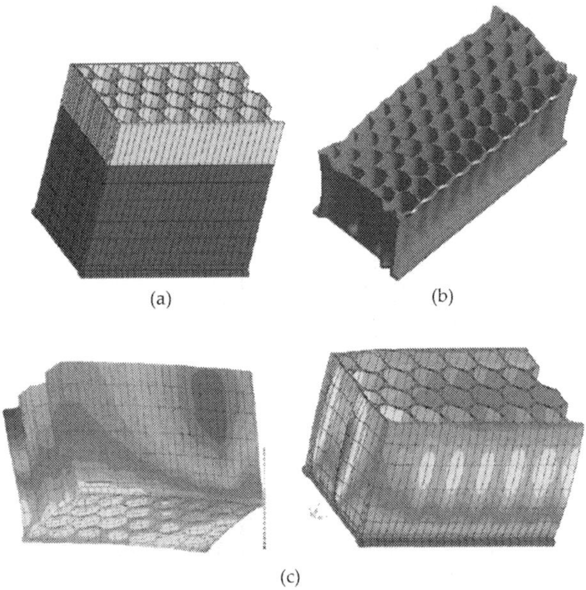

Figure. 5.36: Analysis of a prismatic cellular caisson in Bilbao harbour. (a) Discretiza-tion of half the caisson using 20-noded hexahedra. (b,c) Deformed shape of the caisson and contours of the von Mises stress under lateral wave loading and internal pressure in some cells.

Example 4

Figure 5.37a shows the discretization of the containment building in the Vandellos nuclear power plant in Spain using 20-noded hexahedra. Figure 5.37b shows the amplified deformed shape of the reinforced/pre-stressed concrete structure under a high internal pressure. Colours indicate the displacement modulus contours. Figures 5.37c, d show the damage (fractured) zones for two values of the increasing internal pressure.

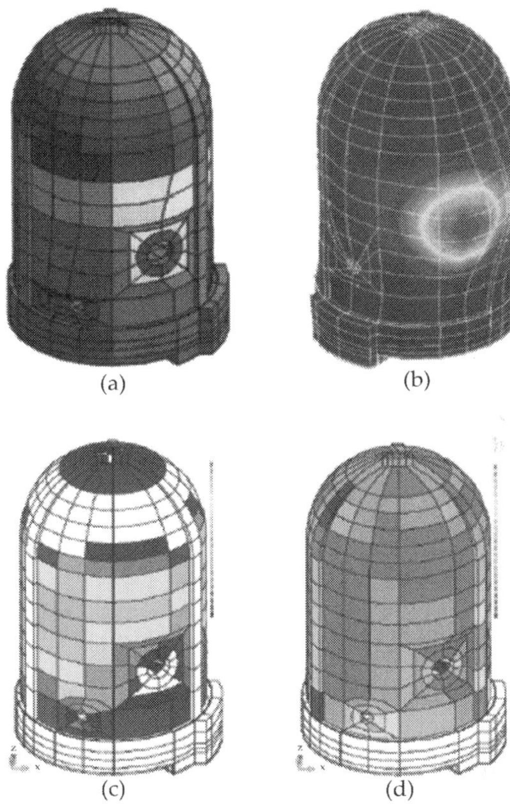

Figure. 5.37: Analysis of a nuclear containment building. (a) Discretization using 20-noded hexahedra. (b) Amplified deformed shape of the structure under internal pressure. (c,d) Damaged (fractured) zones for increasing internal pressure. Colours indicate the normalized value of damage parameter

Example 5

The next example presented is the structural analysis of Barcelona Cathedral. The numerical study was performed using four-noded tetrahedral elements. Figure 5.38 shows the mesh used for the analysis and contours of the displacement modulus under self-weight.

Figure 5.38 shows a similar example for the study of Saint Mark's Basilica in Venice. The figure shows the discretization of the five dome masonry and stone structure using 20-noded hexahedra, the contours of the displacement modulus under self-weight and an underneath view of damage zones in the structure for an increased value of the self-weight.

Figure. 5.38: Analysis of Barcelona Cathedral using 4-noded tetrahedra. Mesh and contours of displacement modulus under self-weight.

Figure 5.39: Analysis of Saint Mark's Basilica in Venice using 20-noded hexahedra. (a) Mesh. (b) Contours of displacement modulus under self-weight. (c) Underneath view of damaged zones for an increased value of the weight.

Example 6

The next examples show the possibilities of 3D solid elements for the study of mechanical engineering components, such as that shown in Figure 5.40a. A mesh of 250 20-noded hexahedra has been used for this analysis. The loading considered is a combination of self-weight and uniform torsion. Figure 5.40b shows the contours of the displacement modulus. The distribution of tension and compression stresses are displayed in Figures 5.40c and d respectively.

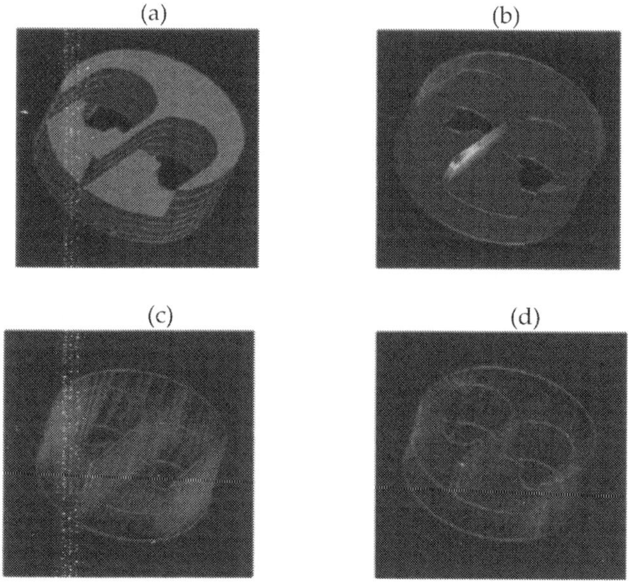

Figure. 5.40: Mechanical component analyzed with 20-noded hexahedra. (a) Geometry and mesh. (b) Contours of displacement modulus under self-weight and torsion loading. (c) Principal tension stresses. (d) Principal compression stresses.

The example shown in Figure 5.41 is the analysis of a mechanical com-ponent using 4-noded tetrahedra. The contours of the von Mises stress under thermal loading are shown as an example of output for this type of problems.

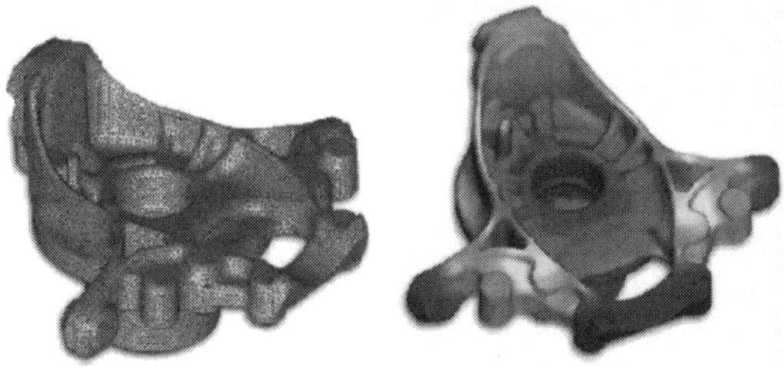

Figure 5.41: Discretization of a mechanical component with 4-noded linear tetrahedra. Contour plot of von Mises stress under thermal loading.

Figure 5.42 finally presents a result of the thermal-mechanical analysis of a crankshaft for an automotive during the solidification and cooling phase of the casting process. Colours show the von Mises stress contours at a certain time of the analysis. A mesh of 4-noded tetrahedral was used to discretize the crankshaft and the surrounding mould.

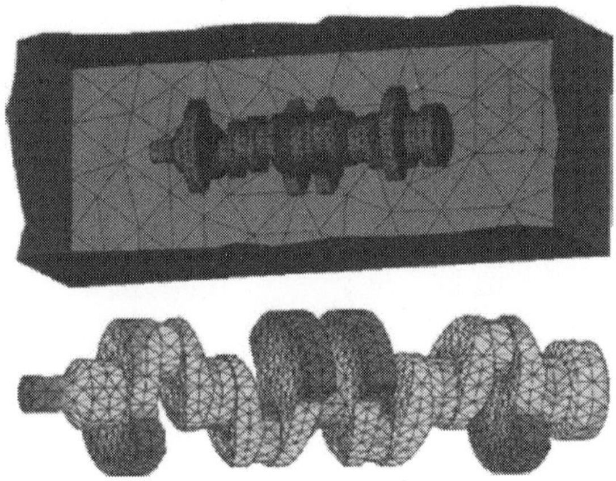

Figure 5.42: Thermal-stress analysis of a crankshaft during the casting process using 4-noded tetrahedra to discretize the part and the mould. Colours indicate the von Mises stress field under thermal loading.

The formulation of 3D solid elements is simple and most concepts are a straightforward extension of those explained for 2D solid elements. 3D solid elements also behave very similarly to the corresponding 2D elements. The 4-noded tetrahedron and the 8-noded hexahedron are the simplest of their respective families, although their accuracy is poorer than that of higher order solid elements. Among these, the 20- and 27-noded hexahedron and the 10-noded tetrahedron are good candidates for practical use.

The examples presented show the difficulties inherent in the discretization and the visualization of numerical results in 3D structural analysis using solid elements. Here the utilization of efficient mesh generators and advanced visualization techniques is essential. Mesh generation is the crucial issue and even more if adaptive mesh refinement procedures involving unstructured meshes of tetrahedra or hexahedra are to be used in an effective manner.

CHAPTER

6

FINITE ELEMENT METHOD IN HEAT TRANSFER

GENERAL PROBLEMS IN HEAT TRANSFER

Problem Statement

Let us consider an isotropic body with temperature-dependent heat transfer. A basic equation of heat transfer has the following form:

$$-\left(\frac{\partial q_x}{\partial x}+\frac{\partial q_y}{\partial y}+\frac{\partial q_z}{\partial z}\right)+Q=\rho c\frac{\partial T}{\partial t}.$$

Here, qx, qy and qz are components of heat flow through the unit area; Q =Q(x,y, z, t) is the inner heat-generation rate per unit volume; ρ is material density; c is heat capacity; T is temperature and t is time. According to Fourier law the components of heat flow can be expressed as follows:

$$q_x = -k\frac{\partial T}{\partial x},$$

$$q_y = -k\frac{\partial T}{\partial y},$$

$$q_z = -k\frac{\partial T}{\partial z},$$

Where k is the thermal-conductivity coefficient of themedia. Substitution of Fourier relations gives the following basic heat transfer equation:

$$\frac{\partial}{\partial x}\left(k\frac{\partial T}{\partial x}\right)+\frac{\partial}{\partial y}\left(k\frac{\partial T}{\partial y}\right)+\frac{\partial}{\partial z}\left(k\frac{\partial T}{\partial z}\right)+Q=\rho c\frac{\partial T}{\partial t}.$$

It is assumed that the boundary conditions can be of the following types:

1 Specified temperature

$$T_s = T_1(x, y, z, t) \text{ on } S_1.$$

2 Specified heat flow

$$q_x n_x + q_y n_y + q_z n_z = -q_s \text{ on } S_2.$$

3 Convection boundary conditions

$$q_x n_x + q_y n_y + q_z n_z = h(T_S - T_e) \text{ on } S_3.$$

4 Radiation

$$q_x n_x + q_y n_y + q_z n_z = \sigma\varepsilon T_S^4 - \alpha q_r \text{ on } S_4.$$

where h is the convection coefficient; Ts is an unknown surface temperature; Te is a convective exchange temperature; σ is the Stefan–Boltzmann constant; ε is the surface emission coefficient; α is the surface absorption coefficient, and qr is the incident radiant heat flow per unit surface area. For transient problems it is necessary to specify an initial temperature field for a body at the time t = 0:

$$T(x, y, z, 0) = T_0(x, y, z)$$

Finite Element Discretization of Heat Transfer Equations

A domain V is divided into finite elements connected at nodes. We shall write all the relations for a finite element. Global equations for the domain can be assembled from finite element equations using connectivity information.

Shape functions Ni are used for interpolation of temperature inside a finite element:

$$T = [N]\{T\}$$

$$[N] = [N_1 N_2 \ldots]$$

$$\{T\} = \{T_1 T_2 \ldots\}$$

Differentiation of the temperature-interpolation equation gives the following interpolation relation for temperature gradients:

$$\begin{Bmatrix} \frac{\partial T}{\partial x} \\ \frac{\partial T}{\partial y} \\ \frac{\partial T}{\partial z} \end{Bmatrix} = \begin{bmatrix} \frac{\partial N_1}{\partial x} & \frac{\partial N_2}{\partial x} & \cdots \\ \frac{\partial N_1}{\partial y} & \frac{\partial N_2}{\partial y} & \cdots \\ \frac{\partial N_1}{\partial z} & \frac{\partial N_2}{\partial z} & \cdots \end{bmatrix} \{T\} = [B]\{T\}.$$

Here, {T} is a vector of temperatures at nodes, [N] is a matrix of shape functions, and [B] is a matrix for temperature-gradient interpolation.

Using the Galerkin method, we can rewrite the basic heat transfer equation in the following form:

$$\int_V \left(\frac{\partial q_x}{\partial x} + \frac{\partial q_y}{\partial y} + \frac{\partial q_z}{\partial z} - Q + \rho c \frac{\partial T}{\partial t} \right) N_i dV = 0.$$

Applying the divergence theorem to the first three terms, we arrive at the relations:

$$\int_V \rho c \frac{\partial T}{\partial t} N_i dV - \int_V \left[\frac{\partial N_i}{\partial x} \; \frac{\partial N_i}{\partial y} \; \frac{\partial N_i}{\partial z} \right] \{q\} dV$$

$$= \int_V Q N_i dV - \int_S \{q\}^{\mathrm{T}} \{n\} N_i dS,$$

$$\{q\}^{\mathrm{T}} = \left[q_x \; q_y \; q_z \right],$$

$$\{n\}^{\mathrm{T}} = \left[n_x \; n_y \; n_z \right],$$

where {n} is an outer normal to the surface of the body. After insertion of boundary conditions into the above equation, the discretized equations are as follows:

$$\int_V \rho c \frac{\partial T}{\partial t} N_i dV - \int_V \left[\frac{\partial N_i}{\partial x} \; \frac{\partial N_i}{\partial y} \; \frac{\partial N_i}{\partial z} \right] \{q\} dV$$

$$= \int_V Q N_i dV - \int_{S_1} \{q\}^{\mathrm{T}} \{n\} N_i dS$$

$$+ \int_{S_2} q_s N_i dS - \int_{S_3} h(T - T_e) N_i dS - \int_{S_4} (\sigma \varepsilon T^4 - \alpha q_r) N_i dS.$$

It is worth noting that

$$\{q\} = -k[B]\{T\}.$$

The discretized finite element equations for heat transfer problems have the following form:

$$[C]\{\dot{T}\} + ([K_c] + [K_h] + [K_r])\{T\}$$

$$= \{R_T\} + \{R_Q\} + \{R_q\} + \{R_h\} + \{R_r\},$$

$$[C] = \int_V \rho c[N]^{\mathrm{T}}[N]dV,$$

$$[K_c] = \int_V k[B]^{\mathrm{T}}[B]dV,$$

$$[K_h] = \int_{S_3} h[N]^{\mathrm{T}}[N]dS,$$

$$[K_r]\{T\} = \int_{S_4} \sigma\varepsilon T^4[N]^{\mathrm{T}}dS,$$

$$\{R_T\} = -\int_{S_1} \{q\}^{\mathrm{T}}\{n\}[N]^{\mathrm{T}}dS,$$

$$\{R_Q\} = \int_V Q[N]^{\mathrm{T}}dV,$$

$$\{R_q\} = \int_{S_2} q_{\mathrm{s}}[N]^{\mathrm{T}}dS,$$

$$\{R_h\} = \int_{S_3} hT_{\mathrm{e}}[N]^{\mathrm{T}}dS,$$

$$\{R_r\} = \int_{S_4} \alpha q_{\mathrm{r}}[N]^{\mathrm{T}}dS.$$

Here, $\{\dot{T}\}$ is a nodal vector of temperature derivatives with respect to time.

Different Type Problems

Equations for different types of problems can be deducted from the above general equation:

Stationary linear problem

$$([K_c]+[K_h])\{T\}=\{R_Q\}+\{R_q\}+\{R_h\}.$$

Stationary nonlinear problem

$$\begin{aligned}&([K_c]+[K_h]+[K_r])\{T\}\\&=\{R_Q(T)\}+\{R_q(T)\}+\{R_h(T)\}+\{R_r(T)\}.\end{aligned}$$

Transient linear problem

$$\begin{aligned}&[C]\{\dot{T}(t)\}+([K_c]+[K_h(t)])\{T(t)\}\\&=\{R_Q(t)\}+\{R_q(t)\}+\{R_h(t)\}.\end{aligned}$$

Transient nonlinear problem

$$\begin{aligned}&[C(T)]\{\dot{T}\}+([K_c(T)]+[K_h(T,t)]+[K_r(T)])\{T\}\\&=\{R_Q(T,t)\}+\{R_q(T,t)\}+\{R_h(T,t)\}+\{R_r(T,t)\}.\end{aligned}$$

Triangular Element

Calculation of element conductivity matrix $[k_c]$ and heat flow vector $\{r_q\}$ is illustrated for a two-dimensional triangular element with three nodes. A simple triangular finite element is shown in Figure 6.1. The temperature distribution T(x,y) inside the triangular element is described by linear interpolation of its nodal values:

$$T(x,y)=N_1(x,y)T_1+N_2(x,y)T_2+N_3(x,y)T_3,$$

$$N_i(x,y)=\alpha_i+\beta_i x+\gamma_i y.$$

Interpolation functions (usually called shape functions) $N_i(x, y)$ should satisfy the following conditions:

$$T(x_i, y_i) = T_i, i = 1, 2, 3.$$

Solution of the above equation system provides expressions for the shape functions:

$$N_i = \frac{1}{2\Delta}(a_i + b_i x + c_i y),$$

$$a_i = x_{i+1}y_{i+2} - x_{i+2}y_{i+1},$$

$$b_i = y_{i+1} - y_{i+2},$$

$$c_i = x_{i+2} - x_{i+1},$$

$$\Delta = \frac{1}{2}(x_2y_3 + x_3y_1 + x_1y_2 - x_2y_1 - x_3y_2 - x_1y_3),$$

where Δ is the element area.

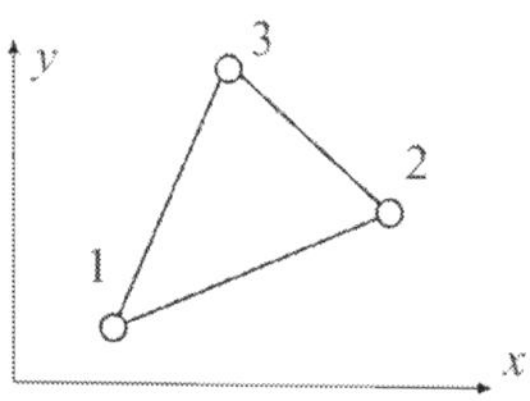

Figure. 6.1: Triangular finite element.

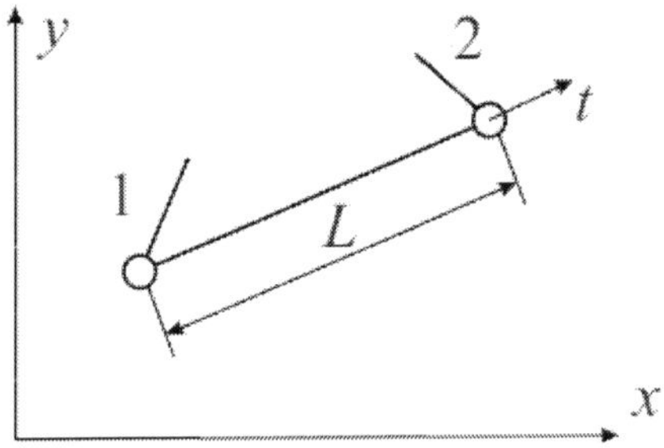

Figure 6.2: Integration along an element side.

The conductivity matrix of the triangular element is determined by integration over element area A (assuming that the element has unit thickness),

$$[k_c] = \int_A k[B]^T [B] dxdy.$$

The temperature differentiation matrix [B] has expression

$$[B] = \begin{bmatrix} \frac{\partial N_1}{\partial x} & \frac{\partial N_2}{\partial x} & \frac{\partial N_3}{\partial x} \\ \frac{\partial N_1}{\partial y} & \frac{\partial N_2}{\partial y} & \frac{\partial N_3}{\partial y} \end{bmatrix} = \frac{1}{2\Delta} \begin{bmatrix} b_1 & b_2 & b_3 \\ c_1 & c_2 & c_3 \end{bmatrix}.$$

Since the temperature differentiation matrix does not depend on coordinates, integration of the conductivity matrix is simple;

$$[k_c] = \frac{k}{4\Delta} \begin{bmatrix} b_1^2 + c_1^2 & b_1 b_2 + c_1 c_2 & b_1 b_3 + c_1 c_3 \\ b_1 b_2 + c_1 c_2 & b_2^2 + c_2^2 & b_2 b_3 + c_2 c_3 \\ b_1 b_3 + c_1 c_3 & b_2 b_3 + c_2 c_3 & b_3^2 + c_3^2 \end{bmatrix}$$

The heat-flow vector $\{r_q\}$ is evaluated by integration over the element side, as shown in Figure 6.2.

$$\{r_q\} = -\int_L q_s[N]^T dL = -\int_0^1 q_s[N_1\ N_2]^T L dt.$$

Here, integration over an element side L is replaced by integration using variable t ranging from 0 to 1. Shape functions N1 and N2 on element side 1–2 can be expressed through t:

$$N_1 = 1 - t, \qquad N_2 = t$$

After integration with substituting integration limits, the heat-flow vector equals

$$\{r_q\} = -q_s \frac{L}{2} \begin{bmatrix} 1 \\ 1 \end{bmatrix}.$$

Element matrices and vectors are calculated for all elements in a mesh and assembled into the global equation system. After application of prescribed temperatures, solution of the global equation system produces temperatures at nodes.

Problems

Obtain shape functions N1, N2, N3 and N4 for the square element shown below.

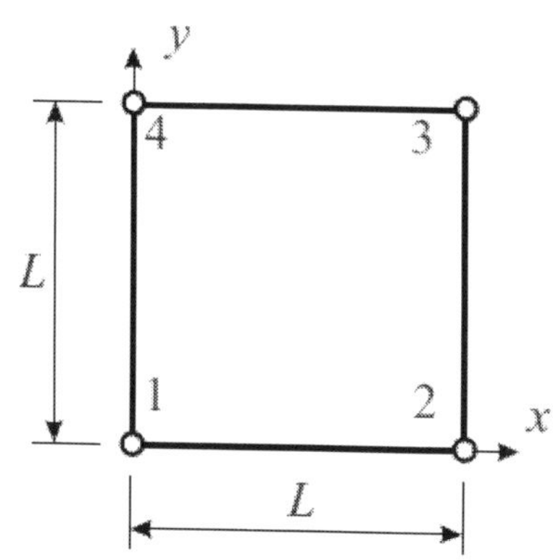

Assume that its size is L=1 and that shape functions can be represented as $N_i = a_1(a_2 + x)(a_3 + y)$.

AN EXAMPLE-CONVECTIVE TEMPERATURE FLOW

Problem Definition

A fluid passes through a domain represented by a pipe having dimensions of 20 x 50 units in the two-dimensional plan, as shown in Figure 1. The fluid flow enters from the left with a uniform velocity V, as illustrated. It flows around a small pipe located at the middle of the domain and having a diameter d = 7.5 units, as shown in Figure 1. This small pipe contains an energy source produces q units of energy per surface area of the pipe per unit time. As the fluid passes through the domain, the velocity of the fluid changes due to the presence of the small pipe at the middle. Also, due to convection phenomenon, heat energy is carried out by the fluid flow through the pipe domain. Thus, the temperature flow field changes by changing the velocity of the fluid flow.

By assuming that the fluid is non-viscous and incompressible, both velocity and temperature fields are independent of each other. As such, the governing partial differential equations that describe the fluid flow and the temperature flow can be solved independently. The fluid flow is solved using the stream function for a specific values of entrance velocities, as will be discussed later. Once the flow is determined, the temperature flow can be established by solving temperature governing equation.

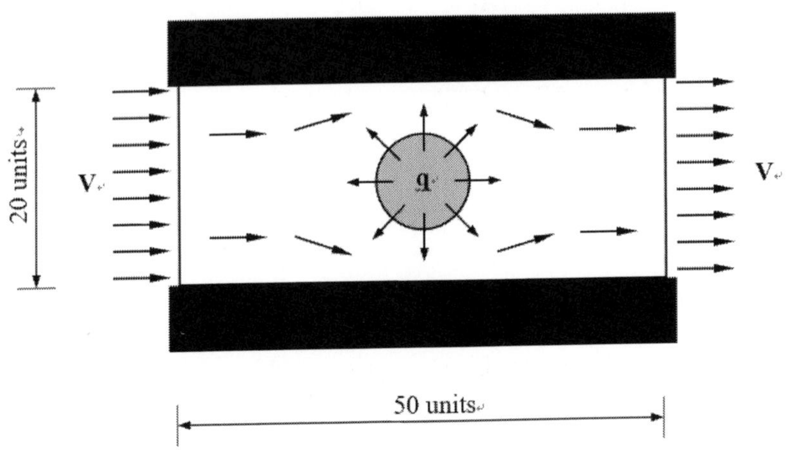

Figure 1: Steady Flow System Definition.

Governing Equations

The governing equation that govern the two-dimensional problem, in general, can be obtained from the relation

$$\frac{\partial}{\partial x}\left(R_x \frac{\partial \text{Ö}}{\partial x}\right)+\frac{\partial}{\partial y}\left(R_y \frac{\partial \text{Ö}}{\partial y}\right)+B_x \frac{\partial \text{Ö}}{\partial x}+B_y \frac{\partial \text{Ö}}{\partial y}+G\text{Ö}+H=0\, in\, \text{Ù}$$

$$\text{Ö}=\text{Ö}^* \quad on \;\; S_1$$

$$R_x \frac{\partial \text{Ö}}{\partial x} n_x + R_y \frac{\partial \text{Ö}}{\partial y} n_y = q^* \quad on \;\; S_2$$

In which the coefficients R_x, R_y, B_x, B_y, G and H are given as functions of x and y. The two-dimensional domain is denoted by Ù . S is the boundary of the domain and S1 and S2 are parts of that boundary and satisfy

$S = S_1 \cup S_2$. n_x and n_y represent the outward normal unit vectors on the boundary.

Governing Equation for Fluid Flow Analysis

Poisson's equation is originated from the general governing partial differential equation and it is adequate to describe a large number of applied problems including the flow of ideal fluids problems. The two governing equations for the flow of ideal fluids are

$$\frac{\partial u_x}{\partial y}-\frac{\partial u_y}{\partial x}=0$$

$$\frac{\partial u_x}{\partial x}+\frac{\partial u_y}{\partial y}=0$$

If we define Ψ such that it identically satisfies the condition for incompressible flow, then we have

$$u_x = +\frac{\partial}{\partial y}\Psi$$

$$u_y = -\frac{\partial}{\partial x}\Psi$$

By substituting equations (4) and (5) into equation (2), the stream function of the fluid flow can be written as follows

$$\frac{\partial^2 \Psi}{\partial x^2} + \frac{\partial^2 \Psi}{\partial y^2} = 0$$

Governing Equation for Temperature Flow Analysis

For steady state two-dimensional convection through a constant-property homogenous fluid, the energy equation

$$\frac{\partial}{\partial x}\left(\rho C_p u_x \varphi - k\frac{\partial \varphi}{\partial x}\right) + \frac{\partial}{\partial y}\left(\rho C_p u_y \varphi - k\frac{\partial \varphi}{\partial y}\right) = 0$$

or in other form

$$\frac{\partial}{\partial x}\left(k\frac{\partial \varphi}{\partial x}\right) + \frac{\partial}{\partial y}\left(k\frac{\partial \varphi}{\partial y}\right) - \rho C_p u_x \frac{\partial \varphi}{\partial x} - \rho C_p u_y \frac{\partial \varphi}{\partial y} = 0$$

Where φ is the temperature, k is the thermal conductivity of the fluid and ρC_p is the heat capacity of the fluid. u_x and u_y denote the velocity components in x and y directions, respectively. Since the time-dependency is beyond the scope of this paper, the thermal conductivity of the fluid will be assumed to be time independent and take a constant value $k = 1.0$. Also, the heat capacity ρC_p is assumed to be equal to 1.0.

Finite Element Modeling

The problem is discretized using a previously developed Matlab code mesh generator "mesh.m" provided. The input data for the mesh generator are the number of generation loops and some geometric coordinates of specific points on each side of each loop. Also, the number of divisions per each side is required to specify the number of elements per each loop. In the current study, six loops are used to generate the mesh of the domain. Thus, the total used number of elements is 560 elements and that corresponds to a total number of nodes 616. A four-node isoparametric quadrilateral element is used to model the domain to maintain the continuity of the degrees of freedom along the edges of the elements. It should be noted, as shown in Figure 2, that the sizes of the elements decreases with getting closer to the small pipe at the middle to ensure the accuracy of the solution at this area of concentrated stresses. The mesh discretization provided in Figure 2 will be used in both the flow and temperature analyses.

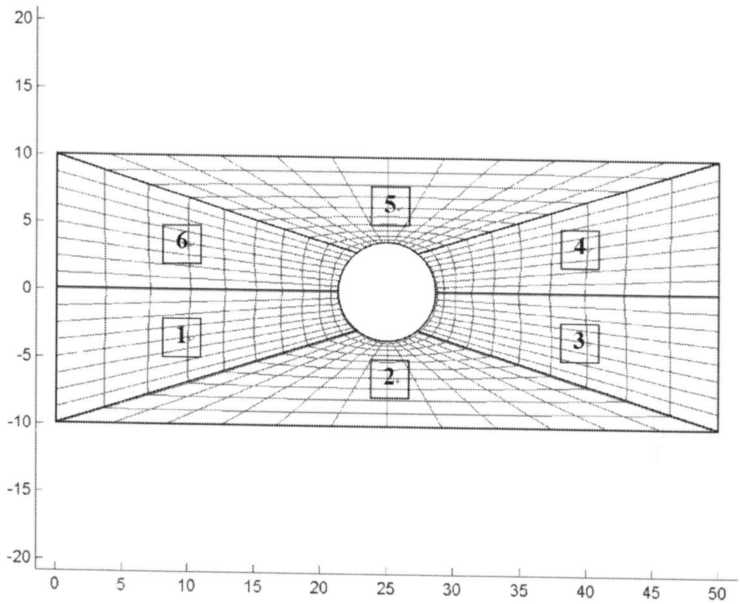

Figure 2: Mesh Discretization of the Domain (560 elements and 616 nodes).

A Matlab user INCLUDE code denoted by NPCODE.m is included in mesh.m code immediately before the output is saved. The purpose of this code is to assign NPcode values to each node. These values have meaning only to the user and are not used in any of the finite element codes. However, the user can use these numbers in any other supplement code. The NPcode values are often used in the INITIAL.m Matlab user INCLUDE code (will be discussed in later section) that set boundary and initial conditions for a particular finite element analysis. The output files of the mesh generator program "mesh.m" are ASCII files; MESHo, NODES and NP. The MESHo file contains the number of nodes, number of elements, and number of nodes per element, respectively. The NODES file includes three columns and number of rows equal to the total number of nodes. The first two columns contain X and Y coordinates of each node, while that last column contains the NPcode value for each node. Finally, NP file includes the connectivity array between the elements.

A problem encountered with the two-dimensional problem analyses is how to number the nodes such that it minimizes the storage needed for the stiffness matrix. The bandwidth of the stiffness matrix depends on the way the nodes have been numbered. Also, the difference between a good numbering scheme and a poor numbering scheme can result in a very large difference in bandwidth requirements. Since finite element equations are related to each other only through common elements, reduction of the bandwidth needs nodes that are connected by common elements be as close in numerical value as possible. A previously developed bandwidth reduction program developed using Matlab code denoted newnum.m is used in the analysis to specify a new numbering scheme that is able to reduce the bandwidth of the stiffness matrix. The ASCII files that are resulted from the mesh.m program are used as input files for the newnum.m program.

The code described in the newnum.m program begins with the specification of one or more nodes as starting nodes. The user specifies these nodes be the first nodes in the new numbering scheme. For example, if five nodes are designated, the line connecting these nodes is referred to as the first wave of nodes (Figure). The second wave consists of all nodes that are linked to nodes in the first wave through common elements. The nodes in the second wave are then given the next consecutive numbers in the new order. This process continues until all nodes have been given new numbers. Finally, all elements should have node numbers that differ by no more than the number of nodes in the longest two consecutive waves, and even less. The output of newnum.m program is an ASCII file denoted NWLD, which contains the new numbering scheme that will be used in the finite element analysis.

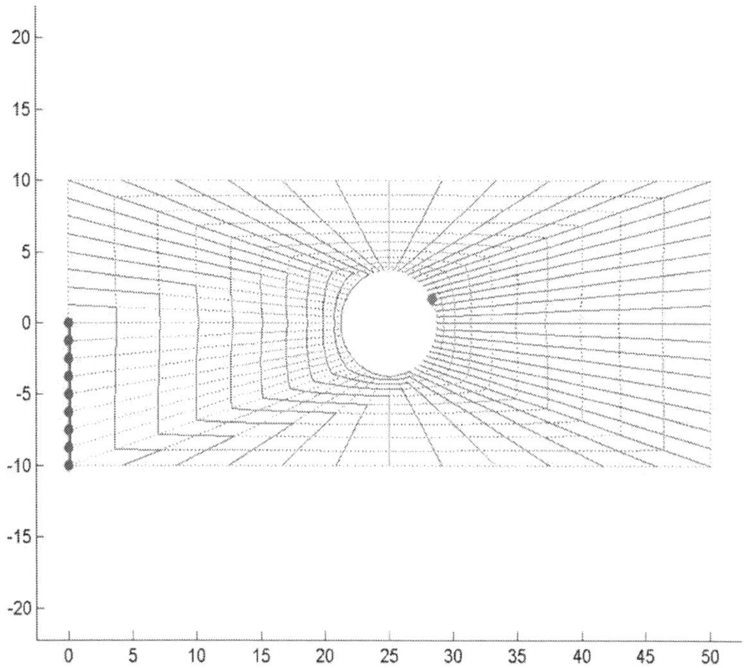

Figure 3: Wave Fronts for New Numbering Scheme (Green is first wave and red dot is last node).

Fluid flow analysis

Both fluid flow and temperature flow can be described by the general equation that governs the two-dimensional problems. However the coefficients of the governing Equation (R_x, R_y, B_x, B_y, G and H) differ from the fluid flow to the temperature flow such that they satisfy the governing equations given in Equations (6) and (8) for fluid flow and temperature flow, respectively. The previously developed Matlab program denoted steady.m is used to conduct the finite element analyses for fluid and temperature flow. The flow chart that describes this program is given in Figure 4

The fluid flow is characterized by the Poisson's equation given by Equation (6) and recalled here for convenience

$$\frac{\partial^2 \Psi}{\partial x^2} + \frac{\partial^2 \Psi}{\partial y^2} = 0$$

in which the coefficients of the general governing equation (1), R_x, R_y, B_x, B_y, G and H, take the following values; $R_x = 1.0$, $R_y = 1.0$, $B_x = 0.0$, $B_y = 0.0$, $G = 0.0$ and $H = 0.0$. These coefficients are incorporated in the user INCLUDE Matlab program denoted COEF.m, which is called from inside the steady.m program to define these coefficients.

The boundary conditions for the fluid flow analysis are illustrated in Figure 5. The x-component of the velocity at the entrance and at the exit is constant. As such, integrating equation (4) for the x-component of the velocity gives the fluid flow conditions at entrance and at exit as

$$\Psi = u_x \, y$$

in which, u_x is constant specific values for the velocity and y is the y-coordinates of the nodes in that boundary. Since, the flow is non-viscous and incompressible, the x-component of the velocity maintains constant at the upper and lower boundaries. Thus, the boundary conditions of fluid flow at upper and lower boundaries can be given by the relations;

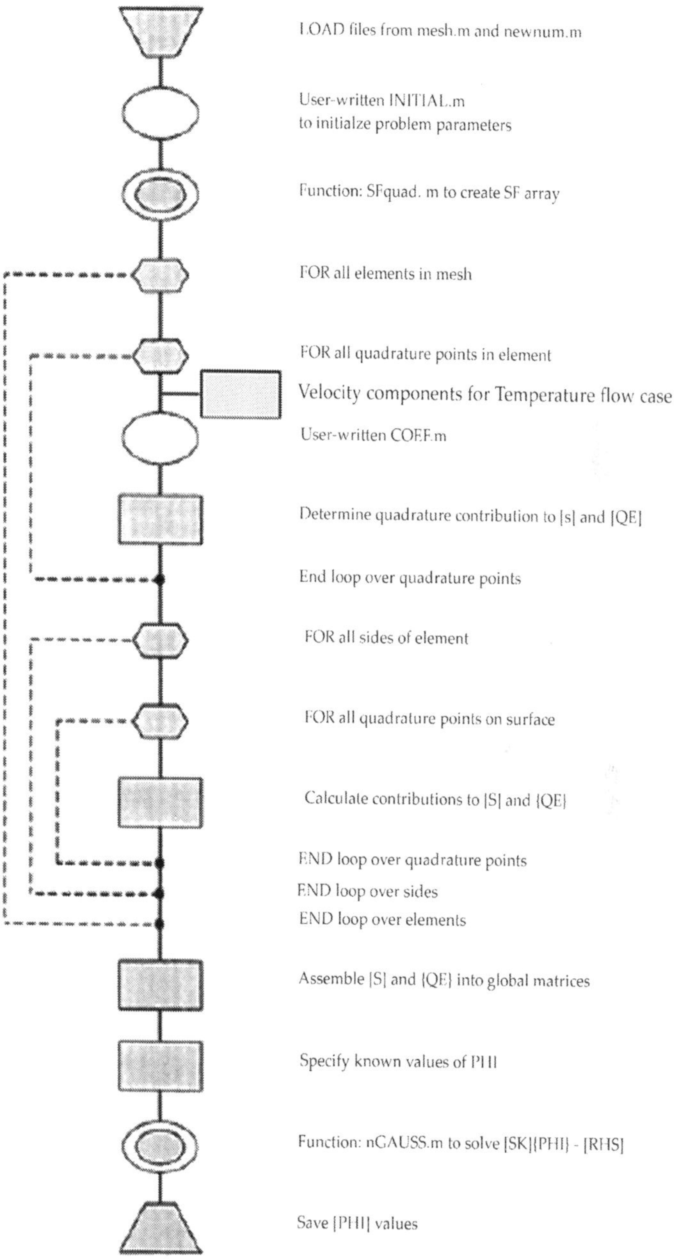

Figure 4: Flow Chart of steady.m Program.

$\Psi = u_x$ (10)

$\Psi = u_x(-10)$

Where the values 10 and –10 are the y-coordinates of the nodes on upper and lower boundaries, respectively. The y-component of the velocity, i.e. $u_y = -\frac{\partial \Psi}{\partial x}$, is constant and equal to zero the at entrance and the exit. Also, to maintain the non-viscosity and incompressibility of the fluid, the fluid flow at the small circular pipe at the middle of the domain is taken equal $\Psi = 0$. These boundary conditions are specified in the INITIAL.m.

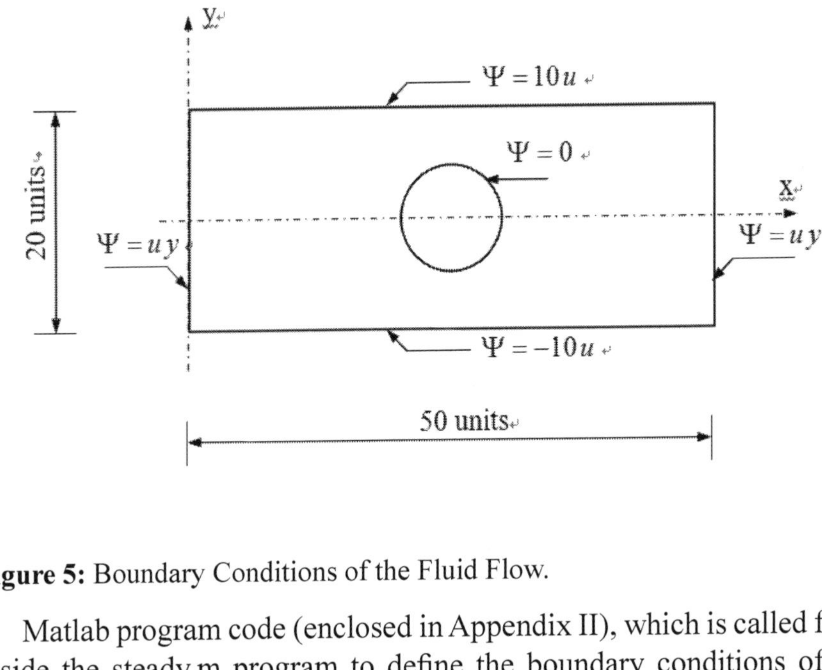

Figure 5: Boundary Conditions of the Fluid Flow.

Matlab program code (enclosed in Appendix II), which is called from inside the steady.m program to define the boundary conditions of the problem. Once the boundary conditions and the coefficients are defined in the steady.m program, the analysis of the fluid flow is determined. Results are obtained for different values of the entrance velocity $u_x =$ 0.0, 0.30, 0.60 and 1.0. Figure 6 shows the flow field of the fluid at the different specified values of the entrance velocity. As illustrated from the results that the flow field is the same for all values of the velocities, however, the absolute values of the flow increases with increasing the entrance velocity. Also, it is noticed that when the velocity $u_x = 0.0$, the flow field is static and the stream function is constant all over the domain.

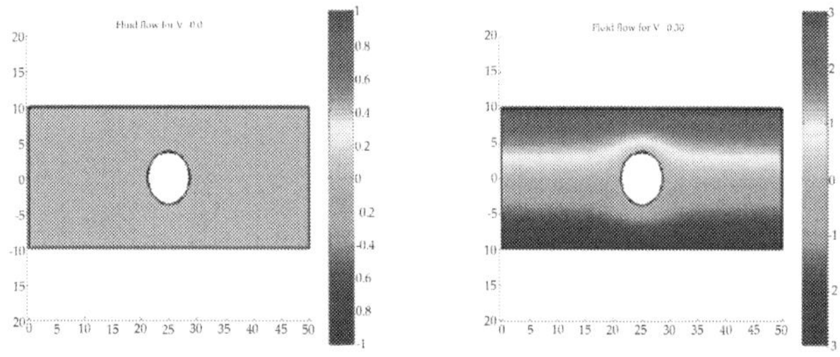

(a) $u_x = 0.0$ (b) $u_x = 0.3$

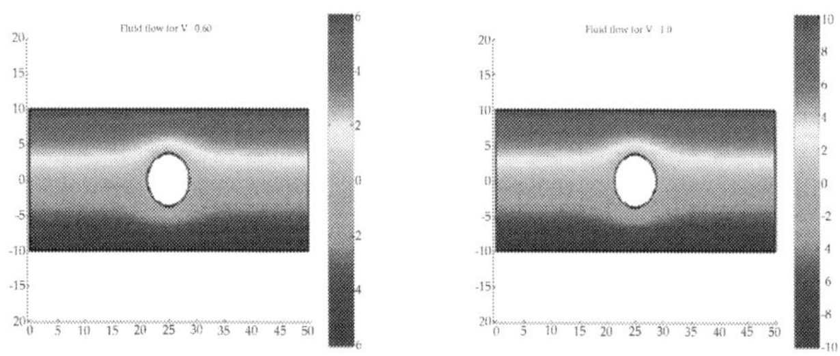

(c) $u_x = 0.6$ (d) $u_x = 1.0$

Figure 6: Fluid Flow Field for Different Values of Entrance Velocity.

Temperature Flow Analysis

The temperature flow is characterized by the energy equation given by Equation (8) and recalled here for convenience

$$\frac{\partial}{\partial x}\left(k\frac{\partial \varphi}{\partial x}\right)+\frac{\partial}{\partial y}\left(k\frac{\partial \varphi}{\partial y}\right)-\rho C_p u_x\frac{\partial \varphi}{\partial x}-\rho C_p u_y\frac{\partial \varphi}{\partial y}=0$$

in which the coefficients of the general governing equation (1), R_x, R_y, B_x, B_y, G and H, take the following values; $R_x = k$, $R_y = k$, $B_x = -\rho C_p u_x$, $B_y = -\rho Cpu_y$, G = 0.0 and H = 0.0. The thermal conductivity k and the heat capacity of the fluid ρC_p are assumed to have unit values. The coefficients B_x and B_y are functions of velocity components u_x and u_y, respectively. Recalling that the velocity components u_x and

u_y are given by using equations (4) and (5) as derivatives of the fluid flow with respect to x and y, respectively. As such, a small Matlab code is incorporated in the steady.m program at the position shown in the flow chart (Figure 4) just before calling the user's INCLUDE file COEF.m to calculate the velocity components u_x and u_y. This Matlab code is illustrated as follows;

```
%    ----------------------------------------
%    Calculating velocity components
%    ----------------------------------------
     for K=1:NNPE;
      NPK=NP(I,K);
      VP060(K,I)=VP06(NPK);
      ux1= ux1+DNDY(K)*VP060(K,I);
      uy1= uy1-DNDX(K)*VP060(K,I);
     end
%    ----------------------------------------
```

in which NNPE is the number of nodes per element, NPK is the global numbering of each node, NP is the array of global numbering of each nodes as function of element number I and local node numbering K. The derivatives of the flow functions with respect to x and y at each node are denoted by DNDY(K) and DNDX(K), respectively. VP06(NPK) is the fluid flow function values, Ψ, resulting from fluid flow analysis for entrance velocity equal u_x= 0.6 given for the global numbering of the nodes. VP060(K,I) is the two dimensional array of the fluid flow function Ψ given at each local node per each element. Once the velocity components u_x and u_y are calculated, the steady.m program call for the user's INCLUDE code COEF.m to read the rest of the coefficient as ;

$R_x = 1$, $R_y = 1$, $B_x = -u_{x1}$, $B_y = -u_{y1}$, G = 0.0 and H = 0.0.

The boundary conditions for the temperature flow analysis are illustrated in Figure 7. The small pipe at the middle is assumed to produce q units of energy per surface area of the pipe per unit time. The energy q is considered equal to unity at the small pipe and equal to zero at exit, lower boundary and upper boundary. The temperature flow φ is considered equal to zero at entrance. These boundary conditions are employed in the INITIAL.m Matlab program code, which is called from inside the steady.m program to define the boundary conditions of the problem. Once the boundary conditions and the coefficients are defined in the steady.m program, the analysis of the temperature flow is ascertained.

Results are obtained for different values of the entrance velocity u_x = 0.0, 0.30, 0.60 and 1.0. Figure 8 shows the flow field of the temperature at the different specified values of the entrance velocity. It is noticed from the results that there is a steep decrease in the temperature field with increasing the entrance velocity. This coincides with the fact that heat transferred by convection is greater than heat transferred by conduction. Also, Figure 8 illustrates that the rate of heat transferred through the domain is increasing by increasing the velocity, which results in decrease of temperature near the heat source in the middle of the domain.

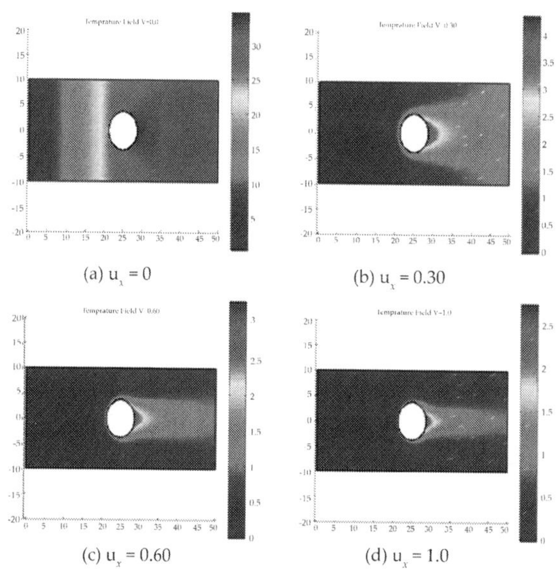

(a) $u_x = 0$ (b) $u_x = 0.30$

(c) $u_x = 0.60$ (d) $u_x = 1.0$

Figure 8: Temperature Flow Field for Different Values of Entrance Velocity.

In this study, a two-dimensional non-viscous incompressible steady flow problem is analysed using the finite element method through solving partial differential equations of the fluid flow. The fluid flow is expressed by partial differential equation (Poisson's equation). While, heat transfer is analyzed using the energy equation. The flow field of that fluid is then used to solve the partial differential equations of temperature flow. The domain of the problem is discretized to a large number of four-node isoparametric elements to assure the accuracy of the solution. The influence of the presence of heat source inside the domain on the temperature flow field is also investigated. Analyses are done by studying both flow field and temperature field for various values of entrance velocity. Results showed that both fluid flow and temperature flow are influenced considerably by changing entrance velocity. Also, there is an obvious effect on the temperature flow field due the presence of an energy source in the middle of the domain.

CHAPTER 7

FINITE ELEMENT METHOD IN MANUFACTURING

The Finite Element Method constitutes a complex process involving a variety of physical phenomena, such as plastic deformation, frictional contact, thermo-mechanical coupling, burr formation mechanisms, etc. Experimental approaches to study manufacturing processes are important but they csan be replaced by FEM analysis with the advantage of saving the time and money required to undertake experimental procedures in the laboratory. However, the experimental validation of FEM analysis in manufacturing processes is also very important. The combinations of FEM analysis and artificial intelligence techniques for modeling and optimization manufacturing process have been used with great success.

METAL CUTTING

There has been a recent increase in the requirement for simulations of metal cutting process, which occurred only when the components of the machining system improved significantly so that the performance of the cutting tool became a decisive factor to further improve productivity, quality, and efficiency of modern manufacturing systems. Several fundamental issues are: the requirement for a physical model, modeling of the behavior of materials during cutting, and verification of the FEM. The discussion of accurate physical models provides the definition and objective of the metal cutting system. The discussion of the behavior of materials during cutting argues that the most common model of behavior, known as the Johnson and Cook model, is unnecessarily overcomplicated. It is also emphasized that verification of the FEM is the final and mandatory stage in metal cutting, with the simple physics-based methods of FEM verification discussed in detail.

Introduction

Although metal cutting, or simply machining, is one of the oldest processes for shaping components in the manufacturing industry and it is widely quoted that 15% of the value of all mechanical components manufactured worldwide is derived from machining operation, machining remains one of the least understood manufacturing operations due to the low predictive ability of machining models despite its obvious economic and technical importance. The metal cutting process is one of the deforming processes where a single-shear plane constitutes the very core of metal cutting theory, and thus this process is thought of primarily as a cutting tool deforming a particular part of the work piece by means of shearing.

A number of cutting theories and FEM models have been developed based on this concept. But these models are not used in the practice of process and tool design. It is partly because tool-manufacturing companies do not want to use them, preferring experience-based methods. Experimental studies in metal cutting are expensive and time-consuming. Moreover, the results are valid only for the experimental conditions used and are highly dependent on the accuracy of the calibration of the experimental equipment and apparatus used. Therefore, it is of the utmost importance to develop a physically sound model of metal cutting and, thus, numerical methods based on the model (for instance, the finite difference method, FEM, boundary element method, etc) with the aim predicting the various outputs and characteristics of the metal cutting process(such as the cutting force, stresses, temperatures, chip geometry, etc.), which can be used in the field appreciate this importance, and therefore, an explanation of the significance of models in this field is required.

In the past. The components of machining system were far from ideal for normal tool performance. Tool specialists (design, manufacturing, and application) were frustrated with old machine tools having spindles that could be rocked by hand; part fixtures that clamped parts differently every times; part materials with inclusions and great scatter in the essential properties; tool holders that could not hold tools without excessive run outs and/or assuring the proper position; starting bushing and bushing plates that had been used for years without replacement; low-concentration coolants which were frequently contaminated; manual with limited range; low dynamic rigidity of machines; etc. As such, the most advanced (and thus expensive) cutting tools performed practically the same (or worse) as the basic tools made in a local tool shop.

Eventually, there was a decrease in the number of tool experts as there was no need for their knowledge. For many years, a stable balance between low-quality (relatively inexpensive tool) and poor machining system characteristics was maintained. Metal cutting research was attributed mainly to university labs and their results were mostly of academic interest rather than a practical significance. It is clear that the metal cutting theory and tool design based on this theory were not requested in practice

It has been gradually changing since the beginning of the 21st century. To address these issues, leading tool and machine manufacturers have developed a number of new products- new tool materials and coating, new cutting inserts and tool designs, new tool holders, powerful precision machines, part fixtures, advanced controllers-which provide a wide range spectrum of information on cutting processes and other aspects of machining and increase the efficiency of machining operations in industry by increasing working speeds, feed rates, tool life, and reliability. For example, modern grades of high-speed steels combined with advanced coating allow high cutting speed that were previously only considered suitable for carbide tools, It is common to use dry hobbing with high-speed steel hobs at 80 m/min so far. Modern grades of polycrystalline diamond (PCD) tool material allow milling, drilling, and reaming of high-silicon alloys at speed of 1000-8000 m/min. Modern grades of carbide tools combined with advanced coating allow machining of alloyed steels at speeds of 300-600 m/min.

Nowadays, machines with great ranges of speed and feed and high-precision spindles, high-pressure coolant supplies and maintenance; high-precision hydraulic and shrink-fit tool holders; tool pre-setting machine (Zollar and Kelch, for example); advanced machine controllers; and many other impovements have become common.

In tool manufacturing system, CNC tool grinders (for example, EWAG, Walter, Star, ANKA, etc) and tool inspection machines (for example, Helickeck, Zoller genius 3) were introduced. In the tool design, 3D solid modeling software package were introduced to help tool designers. New tool materials and advanced grades of the existing tool materials, including nano-coatings, were introduced.

In other words, all the common excuse of inferior tool performance were practically illuminated. The cutting tool manufacturers were pushed to the forefront to show their capability to produce new tools to address the new challenges in metal machining- high productivity rates, low-cost parts, great quality, suitable tool reliability, particularly for unattended manufacturing lines and cells.

The root cause of the failure of the tool manufacturers is the lack of specialists capable of understanding the fundamental principles of metal cutting and tool design. As a result, the design of common tools remains the same as years ago, although modern tool-making technologies, including tool grinding, enable the introduction of entirely new designs to meet modern requirements.

The root cause of the failure of the academic community is that there is no funding available for a seemingly old procedure in which the processes involved should by now be fully elucidated. Specialists were forced to shift their research activities to more fashionable directions, for example, nanotechnology, biomechanics, etc. Gradually, universities abolished courses on metal cutting and tool design, with the result that there is no new pool of specialists.

As a result, there are fewer and fewer professionals in this field (including researchers) and there is less funding for research and development available; these factor cause a number of problems in machining operation efficiency, reliability, and quality in modern manufacturing lines and cells. According to a survey, presented to the American Society of Mechanical Engineering, points out that the lack of knowledge and understanding of metal cutting and cutting tool design is the prime hurdle in increasing the efficiency of automotive plans.

Finite Element Method in metal cutting

There are a number of fundamental issues with proper use of FEM in metal cutting. Authentication of the FEM in metal cutting can be thought of as the process of resolving these issues. To do this. The specialist in the field should understand the nature of these issues, and find the appropriate ways and means to resolve them. The aim of the following section is to discuss the author's view of the most critical issue that must be addressed for the authentication of FEM in metal cutting.

Definition

The system concept in metal cutting was first introduced by Astakhov. According to this concept, the process of metal cutting is defined as a forming process, which takes place in the components of the cutting system that are arranged so that the external fracture of the layer being removed. This fracture occurs due to the combined stress including the continuously changing bending stress causing a cyclic nature of this process. The most important property in metal cutting studies is the system time. The system time was introduced as a new variable in the analysis of the metal cutting system and it was conclusively proven that the relevant properties of the cutting system's components are time dependent. The dynamic interaction of these components take place in the cutting process and cause the cyclic nature of this process.

It follows from the definition that, considered together (the system approach), the following features distinguish metal cutting among other closely related manufacturing process and operations:

Bending moment: the bending moment forms the combined stress in the deformation zone that significantly reduce the resistance of the work material to cutting. As a result, metal cutting is the most energy-efficient material removal process (energy per removed volume accounting for the achieved accuracy) compared to other closely related operations.

Purposeful (micro) fracture: purposeful fracture of the layer being removed under combined stress. The fracture occurs in each successive cycle o chip formation. Although not recognized by the prevailing notion in the metal cutting theory, computational fracture is a part of modern FEM. As such, a great number of separation criteria has been used based on the critical distance, normal failure stress, strain energy, strain energy density, etc. To achieve a reliable FEM simulation, and thus produce reasonable results, the separation criterion must be a sound criterion based on physics of fracture rather than on computation conveniences. Moreover, the energy associated with the formation of new surfaces should also be accounted for.

Stress singularity at the cutting edge: The maximum combined stress does not act at the cutting edge compared to other closely related forming operation. Rather, a (micro) crack forms in front of the cutting edge. As a result, when the cutting system is rigid and the cutting tool is made and run properly, the wear occurs at a certain distance from the cutting edge that enables the accuracy of machining to be maintained over the entire duration of tool time.

Cyclic nature: Metal cutting is inherently a cyclic process. As such, a single chip fragment forms in each chip formation cycle. As a result, considered at the appropriate magnification, the chip structure is not uniform. Rather, it consists of chip fragments and connectors. The frequency of the chip formation process (known also as the chip segmentation frequency) primarily depends on the cutting speed and on the work material. The cutting feed and depth of cut (>1 mm) have very small influence on this frequency.

It follows from the above that the load (the cutting force), temperature field, and other output parameters of process should have cyclic variation. The shape of the chip and its structure should also reflect the cyclic nature of the metal cutting process. Figure7.1 shows an example.

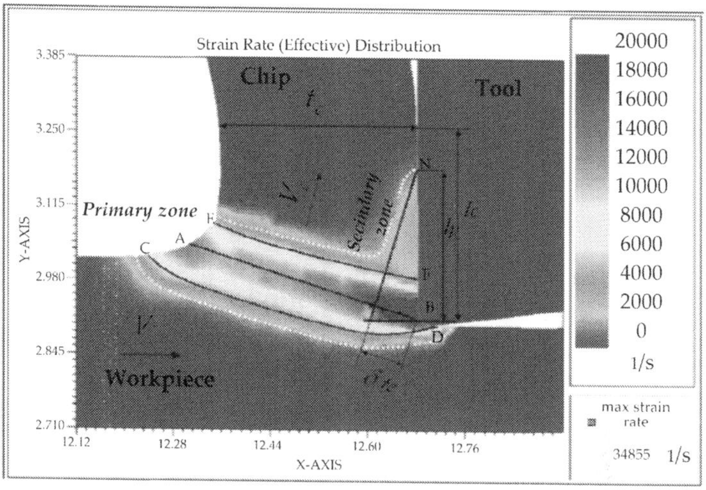

Figure 7.1: Example of FEM simulation by of chip formation in AISI P20 steel and carbide tool, by Tugrul Ozel et al.

Fundamental laws of metal cutting

The first metal cutting law (Makarow's law) formulated by the author in the following form:

For a given combination of tool and work materials, the cutting temperature is referred to as the optimal cutting temperature θ_{opt} , at which the combination of minimum tool wear rate, minimum stabilized cutting force, and the highest quality machined surface are achieved. This temperature is invariant to the way it has been achieved (whether the workpiece was cooled, spre-heated, etc.)

Established initially for longitudinal turning of various work materials, Makarow's law was then experimentally proven for various machining operations. Therefore, the optimum cutting temperature for a given combination of "work material-tool material" should be established and used as the only criterion for the suitability of a particular tool material for a particular work material. This temperature is a physical property and thus does not depend on the intrinsic details of tool design an geometry or on the parameters of a particular test setup. The methodology of determination of the optimum cutting temperature has been previously discussed by the author.

The second metal cutting law, denoted the deformation law, is formulated by the author as:

Plastic deformation of the layer being removed in its transformation into the chip is the greatest nuisance in metal cutting, i.e. although this is required to accomplish the process, it does not add any value to the finished part. Therefore, this energy, which is by far the greatest part of the total energy required by the cutting system, must be considered a waste that should be minimized to achieve higher process efficiency.

The energy of plastic deformation is inversely proportional to the tool life, the quality of the machined surface, and the process efficiency. Therefore, the prime objective of the cutting process design is to reduce this energy to its lowest possible minimum by the proper selection of the tool geometry, tool material, machining regime, coolant, and other design and process parameters.

Material Model

Practically every report on the FEM in metal cutting contains statements on high temperature, high strain, and high strain rate. For example: "Finite element modeling (FEM) of chip formation requires assumption about material flow characteristics and tool-chip contact friction behavior, both at high temperature (1000-1400 K), strain-rate () , and strain (1-4) as encountered during the cutting process"; In extremely high strain-rates (about 1.67×10^5 /s at 500m/min cutting speed and 0.05 mm undeformed chip thickness) and temperatures (about 1400°c) at the chip-tool interface occur in the primary deformation zone and secondary deformation zone, respectively"; "In the cutting process, the deformation of the material in the primary and secondary cutting zones occurs at elevated temperatures and very high strains and strain rates ()".

If, according to the prevailing notion the metal cutting process is accomplished by pure plastic deformation, this process does not have any distinguishing feature among other deforming processes as shown by the author earlier. Therefore, the model of the behavior of work material in metal cutting should be the same as that used in the analysis of other deforming processes. Unfortunately, this is the case in practically all known FEM simulations.

Several material constitutive models are used in the FEM of metal cutting, including rigid plastic, elastic-plastic, viscoplastic, elasto-visco-plastic, etc. These models take into account the high strains and temperatures reportedly found in metal cutting. Among others, the most widely used is the Johnson and Cook model. This is a thermo-elsto-visco-plastic material constitutive model represented as:

$$\sigma_{eq} = (A + B\varepsilon_{eq}^{n})(1 + Cln(\frac{\dot{\varepsilon}_{eq}}{\dot{\varepsilon}_{eq}^{0}}))(1 - (\frac{T - T_0}{T_F - T_0})^m)$$

Where ε_{eq}^{n} is the equivalent plastic strain, $\dot{\varepsilon}_{eq}$ is the equivalent plastic strain rate, $\dot{\varepsilon}_{eq}^{0}$ is the reference equivalent plastic strain rate, (normally), T is the temperature, T_0 is the room temperature, T_F is the melting temperature, and A, B,C, n, and m are constants that depend on the material. These constants are determined through material tests.

Because the strain rates in conventional material tests (tensile, compressive or torsion) are in the range of $10^{-3} - 10^{-1} s^{-1}$, non-conventional material tests, referred to as dynamic tests, are usually preferred by many researchers.

Split Hopkinson pressure bar (SHPB) impact testing is the most common. This test is used to determine the behavior of the material at strain rates up to 10^5. This test can be conducted at elevated temperatures (500°c or even more). However, there are some major problems with the use of the model discussed and the methods of determining of its constants:

Only a few laboratories and specialists in the world can conduct SHPB testing while properly ensuring the condition of dynamic equilibrium.

It is not clear how to correlate the properties of the work materials obtained in SHPB uniaxial impact testing with those in metal cutting with a strong degree of stress triaxiality. As known the eformation and fracture models, as well as the strain at fracture, depend on a given state of stress, which is defined in metal cutting by the tool geometry friction conditions on the tool-chip and tool-workpiece interfaces, machining regime, etc. Even when only the cutting speed varies while all other cutting conditions are constant, the deformation in metal cutting changes significantly, which is reflected in the structure of the chip.

The coefficients of the model are very sensitive to any change in the metallurgical conditions of the work material. According to the existent standards on the composition and properties of various work materials, their properties vary significantly even within the same grade, not to mention heat treatment, grain size, residual stresses, and many other factors. Unfortunately, the many known FEMs of metal cutting used the coefficients of the Johnson and Cook model taken from literature sources. As such, the discussed particularities of a particular work material are not mentioned. For example, Ng et al. used A=950 MPa,B=725 MPa,C=0.015,n=0.375,m=0.625, while Guo et al. used A=792 MPa,B=510 MPa,C=0.014,n=0.26,m=1.03 to describe the behavior of ANSI steel 4340 during cutting.

The model does not account for the formation of cracks and fractures as the model itself does not limit the strain at fracture. In the design of metal-forming operations and tools, the so-called formability limit diagram is used to predict the fracture (failure) of materials under different states of stress; thus, to visualize the limiting plastic strain. Conversely, in metal cutting, the work material is allowed to deform infinitely. As a result, the modeled plastic strains are much higher than those obtained in any known materials testing.

The next concerns to be addressed are the allegedly high strain rate and high temperature of work material in metal cutting, which can potentially affect the properties of the work material relevant to its resistance to cutting, i.e. if the use of the complicated and hard-to-obtain model represented by equation can be justified in principle in the FEM of metal cutting.

The common objective of research work on the application of FEM simulation in metal cutting are appropriate for simulations. For example, common objectives such as “The aim of this paper is to clarify the metal cutting mechanism from the viewpoint of 3D deformation”, “In this paper, a Lagrangian finite element code AdvantEdge was also applied to construct a coupled thermo mechanical finite element model of plain-strain orthogonal metal cutting with continuous chip formation produced by plane-faced uncoated and differently coated carbide tools.”, “The overall objective of this study was to investigate the effect of tool edge geometry on the physics of the cutting process…" are not suitable for simulations.

To obtain more information (mechanics, chemistry, etc.) from the FEM an appropriate amount of information should be put in. This is popularly expressed with the acronym "GIGO", or "garbage in, garbage out." In other words, FEM simulations cannot generate new knowledge. The results of such simulations are new data that (provided the data are correct) help in practical operation and cutting tool design.

Obtaining the stress, temperature, strain, and other outputs, which are also often set as the objective of the study or even of simulations, cannot be considered as the final result because nobody really knows what to do with the data obtained.

Figure 5.1.2.2 shows a typical example where temperatures and plastic strains are simulated. The question of what to do with the results obtained remains unanswered. Because FEM simulation is more an engineering than a scientific computational tool, the results obtained should be used in the design and selection of the proper cutting tools and tool materials, machining regime optimization, and so on.

Unfortunately, the vast majority of the reported FEM simulations does not consider the practical implementations of the results obtained, and continue to report data in a format similar to that shown in Figure 7.2. The only difference is in the colors of user's interface and graphics, which are pretty and "convincing" in the commercial codes.

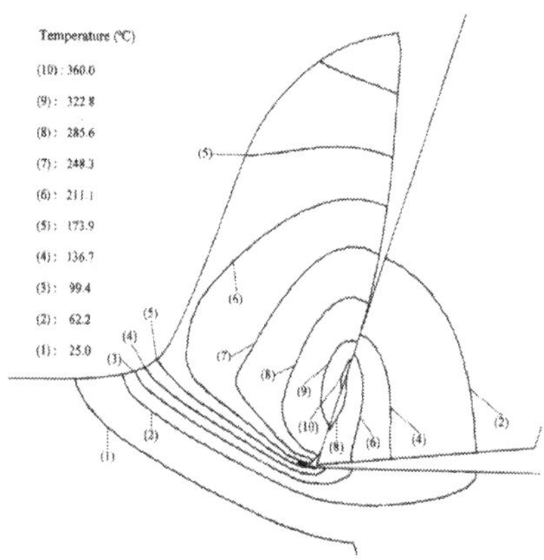

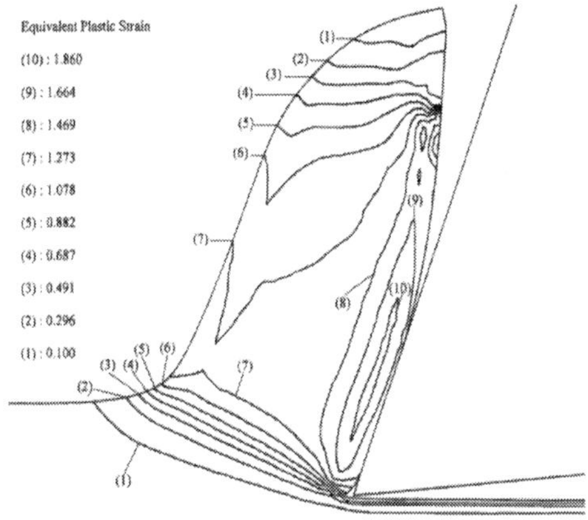

Figure 7.2: Typical results of FEM simulation: (a) temperatures, (b) plastic strains

Therefore, the first and foremost issue to be resolved is the utilization of the results of FEM simulation in practical tool and process design. In other words, these results should be at least of the same level of usefulness as those in the design of metal-forming tools and operations. Thus, the FEM should be an engineering tool incorporated into the design of tools and processes.

Verification of the FEM in metal cutting

Verification of the FEM is the final and mandatory stage of the FEM authentication in metal cutting. This is because the results obtained using FEM strongly depend on the experience and judgment of the engineers involved in the analysis of the problem and definition of a simulation model. The market is overpopulated with FE codes and tools for metal cutting simulations, but the issue of model validity is silently sent aside. How can we seriously propose to eliminate a physical test with a digital simulation if we cannot say that this model has for example, a 95% level of confidence or credibility attached to it?

The best way to validate the model used in FEM simulation is through relevant testing conducted under the same conditions assumed in the model construction. However, this is rarely done because there are a number of problems with adequate metal cutting tests. The experience of scatter in metal cutting experiments, it appears, has now been widely accepted and is no longer under discussion. The uncertainty behind work and tool material properties, working loads, boundary conditions at the tool-chip and tool-workpiece interfaces, and other particularities of the machining system is the main reason why two supposedly identical machining operation will yield different results in terms of tool life, quality of the machined part, and achievable production rate. This fact is well known and yet, surprisingly, engineers are involved in building very detailed models with the intention of capturing the results of one particular test. It is difficult to understand what exactly is detailed in these models, given the fact that the real world is neither detailed, nor accurate, nor precise. Now, as tests are, for very good physical reasons, non-repeatable, this race for accuracy is lost from the very beginning.

The most common method used today to validate (at least, to the first approcimation) a FEM model is a comparison of the cutting force obtained in the modeling and in the cutting test carried out presumably under the same condition as those used in the modeling. Unfortunately, the cutting force cannot be measured with reasonable accuracy, although this fact has never been honestly admitted by the specialists in this field. To appreciate the issue, we should consider the results of the joint program conducted by the College International pour la Recheche en Productique (The International Academy for Production Engineering, http://www.cirp.net (CIRP)) and National Institute of Standards and Technology (NIST) to measure the cutting force in the simplest case of orthogonal cutting. The experiments were carefully prepared (the same batches of the workpiece (steel AISI 1045), tools, etc) under the supervision of NIST and replicated at four of the most advanced metal cutting laboratories in the world. Interestingly, although extraordinary care was taken while performing these experiments, there was significant variation (up to 50%) in the measured cutting force across these four laboratories. Obviously, such accuracy is not acceptable in the validation of the FEM.

The first level of FEM verification includes simple yet often forgotten requirements that should be met by the results of FEM simulation.

Common sense verification:

The shape of the formed chip cannot be changed above the deformation zone. Moreover, the chip cannot became thinner above the deformation zone. The length of the tool chip contact should reduce with the tool rake angle.

The plastic strain gained by the chip in the deformation zone cannot be changed. For example, plastic strain mysteriously decreases reaching the level of the virgin work material. Unfortunately, this is one of the most common mistakes.

Besides some very specific cases, the maximum temperature should be found approximately in middle of the tool-chip interface.

The cutting wedge that represents the tool is not infinitely rigid as assumed in FEM simulations. Rather, it is subjected to elastic deformations at the tool-workpiece interface. Unfortunately, this fact was completely missed in FEM simulation of metal cutting. Otherwise, these simulations could predict the simulation of metal cutting. Otherwise, these simulations could predict the appearance of the so-called tangential location of the cutting insert that revolutionized cutting tool performance by doubling tool productivity.

Geometrical and metallographic similarities:

Geometrical and metallographic similarities are used at the first and the basic or "must" level of verification of the FEM for general purpose, simple FEM codes, for example those invluded in solid modeling CAD programs.

The simplest way to verify the FEM is to ensure the geometrical and metallographic similarities of the model and test results. Figure 7.3 shows an example where four basic parameters of the continuous-fragmentary chip are to be compared:

The maximum chip width of the modeled, W_{max-m} and experimentally-obtained chip, W_{max-e};

The minimum chip widths, W_{min-m} and $W_{min-c;}$

The average pinch of the serrations, p_m and p_e;

The angle of maximum deformation, β_m and β_e.

It is obvious that these parameters can hardly be the same due to the assumptions made in modeling and imperfections of real machining.

Therefore, sound engineering judgment of the research team should help to establish the tolerable degree of similarity.

The choice of parameters to be compared also depends on the chip type and on the research objective. For example, if FEM simulations are to study the influence of the cutting speed then the model should be able to yield the same change in the chip structure/type with the cutting speed as obtained experimentally.

Deformation similarity:

Deforming similarity is used for the verification of FEM models used to optimize the cutting process, i.e. to reduce plastic deformation according to the second law of metal cutting. Revealing the influence of the parameters of the cutting tool (the tool geometry, materials, edge preparation, surface finish, coating, etc.), physical and metallurgical state of the work material, and other important aspects of a cutting operation on the total amount of plastic deformation and minimizing this plastic deformation by the optimization of the parameters is the prime objective.

The similarity theory adopted for metal cutting provides great help in establishing the deformation similarity between FEM simulations and the experimental results. Two similarity criteria, namely, the chip compression ratio and the Poletica criterion are of prime importance in this respect.

The chip compression ratio (hereafter, CCR) ensures similarity of deformation processes in metal cutting. Because the energy spent on plastic deformation of a wide variety of ductile work material is within the range of 70-80% of the total energy, this similarity criterion is probably the most important to verify the validity of a FEM. CCR ζ is determined as the ratio of the length of the cut (L_1) to the corresponding length of the chip (L_c) or the ratio of the chip thickness (t_2) to the uncut chip thickness (t_1), or the ratio of the cutting speed (v) i.e.

$$\zeta = \frac{L_1}{L_2} = \frac{t_2}{t_1} = \frac{v}{v_1}$$

Comparing the simulated and experimentally obtained CCRs, we can determine the validity of FEM with a great precision because this parameter is probably the easiest outcome of the cutting process and of simulation to measure with sufficient accuracy. Moreover, because for many common work materials, the CCR is already established as a functioning of major process parameters, the "express" analysis of the validity of the simulation can be used. For example, the CCR for turning 316L stainless steel is ζ=2.5-3.5, therefore the average chip thickness in FEM simulations should be 2.5-3.5 times greater that the uncut chip thickness. Using this knowledge and comparing the simulated chip thickness to the uncut chip thickness shown in Figure we can conclude that the simulated result is not correct as it does not pass even a simple express analysis of deformation similarity.

When a practical cutting operation (for example, longitudinal turning) is used to verify a FEM simulation made for orthogonal cutting, few relatively simple adjustments should be made to obtain the correct results. These include the determination of the true uncut chip thickness and taking into account the real radius of the cutting edge. These adjustments, however, are relatively simple and self-evident so that the verification can be carried out even at the shop-floor level. The simple practical methods of experimental determination of the chip compression ratio ζ, for common machining operations have been discussed.

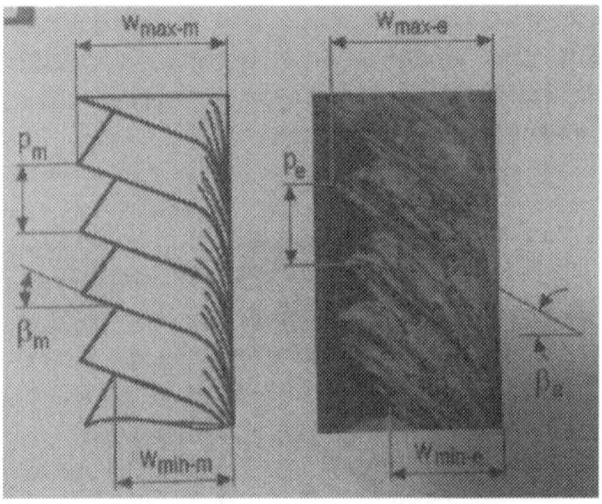

Figure 7.3: Comparison of the basic parameters of the model and experimentally obtained chips.

The Poletica criterion (Po-criterion) is used to assess the length of the tool-chip interface. It is defined as the ratio of the contact length L_c, to the uncut chip thickness t_1

$$P_O = \frac{l_c}{t_1}$$

Although the Po-criterion can be measured using a cutting test, it was found that for a wide variety of work material this criterion can be calculated through CCR ζ, as

$$P_O = \zeta^{k_t}$$

Where k_r=1.5 when $\zeta<4$ and k_r=1.3 when $\zeta\geq4$.

In metal cutting, the tool-chip contact length known as the length of the tool-chip interface determines major tribological conditions at this interface such as temperatures, stresses, tool wear, etc. Moreover, all the energy required by the cutting system for chip removal passes through this interface. Therefore, the similarity of the modeled and experimentally obtained Po-criteria assures the similarity of these phenomena in FEM and in reality.

Thermodynamic similarity:

Thermodynamic similarity is important when FEM simulations are used to achieve the maximum tool life under a desired productivity and/or efficiency of a machining operation utilizing the power of the first metal cutting law. Unfortunately, a simple way to verify this similarity has not yet been developed. Measuring the temperature field using thermocouples is complicated and not accurate. Infrared thermography is increasingly used as an emerging experimental technique in metal cutting for various temperature measurements. The major advantage of infrared thermography is that it does not interfere with the cutting process. Over recent years, low-cost camera solutions were applied to the determination of the temperature distribution on the tool and chip in orthogonal cutting. However, the most severe disadvantage of this technique is that it can only be applied to determine the temperature of the surface if this surface is available for direct observation by the infrared camera. As a result, the maximum contact temperature and its exact location with respect to the cutting edge may not be accurately determined. Even measuring the temperature field of the chip-free surface presents a problem because this surface is not flat at micro (serrations) and macro (chip curing) levels.

HIGH-SPEED MACHINING

Introduction

Temperature and heat distribution prediction in metal cutting are the key factors that facilitate in predicting wear mechanisms and tool life during the machining process. If these factors are not controlled, premature tool failure can occur and subsequently affect machining efficiency. These problems have to be targeted during the design phase of cutting tools. During this phase, numerical-based tools can be effectively and conveniently utilized to predict the factors that are critical to the machining process. Amongst these, the finite element method (FEM) has been widely used to estimate the performance of cutting tools.

Currently, a highly competitive environments driving the developments in the manufacturing industry. Manufacturers are making efforts to speed up their manufacturing processes with improved quality and timely delivery. Like other manufacturing processes, machining has also been the subject of much attention. In this area the tendency is towards a high material removal rate with a high degree of accuracy. High-speed Machining (hereafter, HSM) has attracted the interest of manufacturers due to its favorable response to complex machining and very close tolerance. HSM provides high productivity, better surface quality (R_a=0.2μm) and appropriate accuracy of machined parts and hence overall competitiveness.

The main application of HSM includes the machining of aluminum and steel parts in automotive industry and the finishing of hardened mold material in the die/mold industry. Hardened mold components are normally machined from alloy steels with hardness between 60-65 HRC and the machining of long/thin aluminum parts, which are generally used in the automotive and aerospace industries.

The distinction between conventional and HSM can be defined using different criteria including: magnitude of the cutting speed, revolution of the spindle, dynamic behavior, material removal rate, and tool coating grades. However, the classification developed by Schulz and Moriwaki is the most popular and is based on the material to be machined. According to this classification each material (to be machined) has its own range of conventional, transitional an HSM regimes. The high-speed cutting range of steel is the conventional speed range of steel is the high-speed cutting range of titanium, and the conventional cutting range of titanium is high-speed for nickel-based alloys. It means that with increasing hardness of the material, the cutting speed range reduces and with increasing softness of the material, the cutting speed range increases. Therefore, this classification is widely used and differentiates the different cutting speed regimes for different workpiece materials.

In HSM, the worekpiece material is deformed under extremely intense conditions in a very narrow band. Due to high deformation and deformation rate, the workpiece material suffers severe plastic deformation in the primary deformation zone. It has been reported that the strain rate during HSM process can be elevated up to , which is considerably higher than any other production process or conventional material testing methods.

The cutting speed has a dominating effect on tool-chip interface temperature and heat transfer. The short contact time minimizes the strain hardening and heat-affected zone in the workpiece material. Therefore, most of the heat generated at the shear plane flows into the chip rather into the workpiece and because of this, the HSM process is considered nearly adiabatic.

Due to an extremely narrow heat-affected zone along the shear plane and high temperature of the order of approximately1000, the benefits of the cutting fluid are not so significant. Due to high rotational speeds and short interaction time cutting fluid cannot access to the narrow zone. Dudzinski et al. have argued that the use of coolant in machining at high cutting temperatures can lead to thermal shocks, which have detrimental effects on tool life. For this reason, HSM is generally described as dry machining. Although HSM has some major advantages, it also has some disadvantages. According to Fallboehmer et al., the disadvantages using HSM are high tool wear, the need for special and expensive machine tools with advanced cutting tool materials and coatings. Therefore, concurrent with an increase in HSM applications, there is a need to investigate and research new cutting tool materials and geometry, improved coating compounds and optimized cutting conditions.

Finite Element Method in HSM

To study the machining process, analytical and numerical techniques are frequently used to support the experiments, Due to the highly complex nature of the thermomechanical process (in machining) with large deformation, high strain rates, excessive temperatures, and complex loading conditions, analytical modeling is an issue. Most of the analytical models are based on simplifying assumptions, which do not necessarily hold true under certain conditions. Without simplifying assumptions and appropriate boundary conditions, the analytical approach renders a complex set of mathematical equations for which a solution is no longer possible.

Due to these complexities and limitations, numerical tools such as the FEM have been used to study metal machining. The following are some advantages of using FEMs in machining studies:

Strain, strain rate, and temperature-dependent material properties can be assigned;

Various friction boundary conditions between the chip and the tool can be modeled;

Geometric non-linearities, such as free boundaries of the chip surface, can be easily dealt with;

Difficult to measure variables, such as chip geometry, local stresses, and heat generation and distribution, can be conveniently obtained.

Although FEM has many advantages over the analytical methods, it is still not fully capable of simulating a real machining operation. It is due to the complexity and diversity of the cutting processes, a single process model or simulation strategy cannot be applied to all materials and cutting conditions.

Pioneered by Tay et al., who developed a 2D FEM model for orthogonal turning , they used the FEM to estimate the temperature fields generated in the workpiece, the chip, and the tool durin orthogonal machining Usui and Shirakashi used a stready-state, elastic-plastic material model assuming rate-independent deformation behavior. For chip separation, however, they used crack initiation at the cutting edge, which contradicts the established plastic deformation theory of the machining process under steady-state conditions.

FEM modeling techniques can be effectively used to study the machining process. However, there are still several issues related to the real machining process, which need to be incorporated in FEM simulations. These issues have so far not been highlighted carefully and hence FEM modeling of machining processes modeling of machining processes remains critical. These issues are material flow properties, chip separation, and friction models. Soo and Aspinwall argued that simulation of machining will only be reliable if the criteria used for modeling reflect the physical mechanisms. Thus, FEM simulation of machining processes is still imprecise and requires further modeling refinement.

FE transient heat transfer analysis

This is an example using commercially FE software, Abaqus/Standard 6.10-1 to perform conduction heat transfer transient analysis of the cutting tool with multilayer coating and restricted contact length groove type system. Due to a significantly high modulus of elasticity, the cutting tool was modeled as a perfectly rigid body. A four-node tetrahedral heat transfer DC3D4 element was used to model the whole assembly of the cutting tool. It can be seen from Figure that the complete tool has been meshed. This is because during the cutting process heat is distributed into the shim and the tool holder, and therefore, the computational domain cannot be considered as cutting insert only. The FE mesh was refined in the regions where large temperature gradients were expected, i.e. at the tool-chip contact area. Details of the insert holding screw were neglected. Also chip breaker profile geometry was made only at one edge due to the modeling limitation of the analysis package.

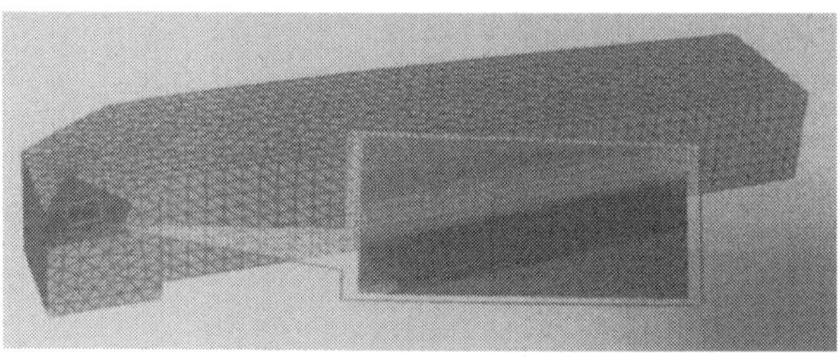

Figure 7.4: FE mesh for cutting tool assembly in Abaqus/Standard 6.10-1

The transient heat transfer problem is solved/governed by the following partial different equation:

$$\frac{\partial^2 \theta}{\partial x^2} + \frac{\partial^2 \theta}{\partial y^2} + \frac{\partial^2 \theta}{\partial z^2} + \frac{q(x, y, z, t)}{\lambda_T} = \frac{1}{\alpha_T}\frac{\partial \theta}{\partial t}$$

Where α_T and λ_T are the thermal diffusivity and thermal conductivity, respectively, and θ is the temperature. As the thermal diffusivity and thermal conductivity are functions of temperature, temperature-dependent thermal properties were used, as given by the table.

Table Temperature dependent mechanical and thermal properties of cutting tool system

		Tool holder AISI/ SAE 1045	*WC*	*TiN*	*TICN*	Al_2O_3
Young's modulus, E (GPa) Density, ρ (kg/ m^3) Poisson's ratio ν 7844 0.30		*207*	*534*	*250*	*355*	*340*
		11900	*5420*	*4180*	*3780*	
		0.22	*0.25*	*0.20*	*0.23*	
Thermal conductivity, λ T at temperatures (°c)	*100*	*50.70*	*40.15*	*21.00*	*29.00*	*17.00*
	200	*48.20*	*44.35*	*21.47*	*29.90*	*14.10*
	300	*45.30*	*48.55*	*22.00*	*30.60*	*12.50*
	400	*41.90*	*52.75*	*22.52*	*31.50*	*10.80*
	500	*38.10*	*56.95*	*23.00*	*32.00*	*8.75*
	600	*33.90*	*61.15*	*23.72*	*33.00*	*7.50*
	700	*30.10*	*65.35*	*24.38*	*33.50*	*6.50*
	800	*24.70*	*69.55*	*25.01*	*34.50*	*6.00*
	900	*25.75*	*73.75*	*25.50*	*35.00*	*5.50*

Specific heat, at temperatures (°c)	*100*	*470.40*	*346.01*	*702.60*	*1030*	*903*
	200	*520.80*	*358.01*	*752.70*	*1020*	*1022*
	300	*571.20*	*370.01*	*783.40*	*1040*	*1089*
	400	*621.60*	*382.01*	*801.16*	*1070*	*1139*
	500	*672.60*	*394.01*	*818.90*	*1120*	*1176*
	600	*722.80*	*406.01*	*833.46*	*1260*	*1202*
	700	*772.80*	*418.01*	*846.39*	*1350*	*1220*
	800	*823,20*	*430.01*	*856.00*	*1660*	*1237*
	900	*873.60*	*442.01*	*857.60*	*1810*	*1252*

The boundary conditions applied for FE thermal analysis of the cutting tool are:

The initial temperature of the model was set at room temperature, i.e.25°c;

The far end surface of the tool holder, which is distant from the cutting zone, was assumed to be at room temperature during simulation (i.e.25°c);

The non-interactive surfaces of the model, which are exposed to air except tool-chip contact area, heat loss due to convection was considered as;

The coated insert system, the shim, and tool holder were assumed to be in perfect contact;

The heat flux applied was assumed to be constant across the contact area, i.e. in the direction of the depth of cut (a_p).

The fraction of heat R^T entering the cutting tool insert was estimated using a hybrid FE and experimental methodology. Initially 100% heat flux, uniform and non-uniform, where appropriate, was applied on the tool-chip contact area. The heat flux was then reduced until temperatures at six points and temperature trend on each path were matched with experimentally measured temperatures on the rake face. The novelty of this approach in this study is the validation of the model, by matching path temperatures for groove profiled cutting tools instead single point matching.

Verification

Experiments

The hybrid methodology to evaluate the heat distribution into the cutting tool (in this study) needs experimental data as an input to FEM modeling. Therefore for this purpose, cutting tests (nearly orthogonal machining) were conducted using semi-automated Dean Smith and Grace Lathe machine. A Kislter three-component piezoeletric dynamometer type 9263 was used to measure the cutting forces (i.e. F_C and F_f). The amplified signals were recorded on the PC using the data-acquisition software system. Commercially available cutting inserts from Sandvik Coromant (having ISO specification), TCMT 16T308, restricted contact length groove type were used. The coating grade was GC 4215 (ISO P15-CVD coated carbide). The cutting inserts were multilayer coated having a tungsten carbide substrate with TiCN/Al_2O_3 coatings on the rake face and TiCN/Al_2O_3/TiN on the flank face. The coating layout on the (a) rank face, and, (b) flank face is shown in Figure, respectively. The thicknesses of the coating were measured using an SEM. The thicknesses of TiCN, Al_2O_3 and TiN were 7, 7, and 2 μm, repectively. The tool holder was a right hand Sandvik STGCR 2020K-16 with a 0° rake angle and 7° clearenc angle.

The workpiece material AISI/SAE 4140 (low-carbon high-tensile alloy) was bored to make a tube of thickness 2 mm with an external diameter of 200 mm. The cutting speeds were set on the basis of available spindle speeds on the lathe. The cutting tests were conducted without coolant (dry machining) with seven different cutting speeds (all are high speeds), i.e. 314, 395, 446, 565, 628, 785 and 879 m/min. For each cutting speed a new cutting edge was used. The feed and depth of cut were kept constant at 0.15 mm/rev and 2 mm, repectively.

Temperature measurement

Temperature profiles generated on the tool rake face during the cutting tests are important to validate FE thermal modeling. For this purpose, infrared thermal imaging was used in obtaining the temperature during cutting on the rake face. An infrared thermal imager FLIRCAM SC3000 with ThermaCAM Researcher package was used. This camera is a long-wave, self-cooling analysis system with a cool down time of <6 min.

This camera system can measure the temperatures from -20 to 2000°c. Up to 150°c, the measuring accuracy is ±2°c (or ±2%). With the ThermaCAM Researcher package this camera can acquire images and data at high rates, up to 750 Hz PAL/900Hz NTSC. According to Abukhshim et al. this camera package allows extensive analysis of highly dynamic objects and events typically found in metal machining research applications.

The camera was mounted on the machine turret and placed in such a way that the tool rake face would be captured clearly. As the cutting inserts were restricted contact length groove type, there was a possibility of a segmented chip to be produced with higher velocity. Therefore, the camera was positioned at 40cm away from the cutting region to avoid any damage to the camera.

The thermal images captured by the camera were analyzed using ThermaCAM Researcher package. The temperatures were recorded using thermal maps during cutting on two different paths (Path-1 and Path-2), as shown in Figure, at six equally spaced points on each path. These paths were appropriately selected on either side of the chip flow zone, which were prominent during the cutting and were not obstructed by the chip flow. The groove profile of the chip breaker was avoided for the consideration of the paths due to its non-horizontal geometry (relative to the whole cutting tool system), i.e. groove profile lies in more than one plane.

Accurate radiation thermometry, including temperature measurement. According to Sievers an dRavindra, et al, emissivity plays a key role in determining the actual temperatures of the metal and alloys. The surface emissivity can be influenced by several factors, including material composition, surface texture, and the calibration method. Considering this fact, another experiment was carried out in order find the appropriate thermal emissicity of the tool rake face rather than using the emissivity value from previously reported literature (forAl_2O_3).

For this purpose, the cutting tool inserts samples were heated to temperatures in the range of 100-1000□. Thermal imaging camera along with ThermaCAM Researcher package was used to measure the temperatures. Emissivity value in the ThermaCAM Research package was adjusted to match the temperature of a particular point with both thermocouple and pyrometer (two colored) readings. At each temperature level data of 20 emissivity values were obtained and the average of these (in the range is 0.4-0.65) was used for the camera calibration of the cutting test.

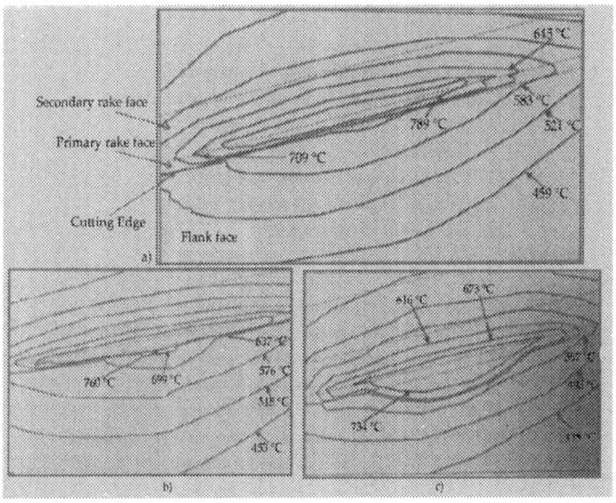

Figure 7.5: Finite element simulation temperature (plots) distribution at cutting speeds of a) 314 m/min b) 556 m/min c) 879 m/min

Heat partition into the cutting tool

This part discusses the results of the hybrid methodology adapted to estimate the amount of heat going into the tool for groove-profiled restricted contact length. Figure shows the simulated temperature distribution on the cutting tool at cutting speeds of 314, 565 and 879 m/min. A uniform and non-uniform heat (where appropriate) was applied at the tool-chip contact area. The tool has $TiCN/Al_2O_3$ coatings on the rake face and $TiCN/Al_2O_3/TiN$ on the flank side. The simulated results shows high temperatures in a very narrow zone near the cutting edge.

This was expected as at higher speeds and contact length shrinks to the cutting edge and hence produces higher temperatures. Note from Figure that at lower cutting speed (relative to 556 and 879 m/min), the higher temperatures can be observed in the tool-chip contact zone. In the case of a cutting speed of 556 m/min, as shown in Figure, there is a shift in the location for the higher temperature, i.e. near the cutting edge. At 879 m/min (Figure) high-temperature profiles cross the cutting edge towards the flank face. This is due to the shift in the contact area towards the cutting edge at higher speeds, instead of heat spreading on the rake face, it starts to extend on the flank face. This is a major cause of the flank wear at higher cutting speeds and can be attributed as one of disadvantages of HSM.

For the prediction of heat partition into the cutting tool a hybrid FE approach and experimental data have been used. The validation of the model has been done by comparing the FE-simulated temperatures to the experimentally measured temperatures on the preselected paths (six points on each path, as shown in Figure). Experimentally measured and the FE-simulated temperatures along path 1 and 2 for the cutting speeds 314, 556, and 879 m/min are shown in Figure, respectively.

For FE simulation, temperature at six points (on each path) exactly at the same location as the experimental points and also the temperature trends for each path (11 points) were ploted (as shown in Figure), to match the temperature at particular points and the temperature trend, respectively.

The experimental temperature measurements using the ThermaCAM were repeated and mean values of the temperatures at six equally spaced points on each path were used for temperature matching. In general FE simulations overestimated temperatures at lower cutting speeds, i.e. between 314 to 446 m/min and at higher cutting speeds, i.e. 565 to 879 m/min, the temperatures are underestimated.

The difference between the temperature trends (of both paths) of FE and experimental results at lower cutting speeds (relative to higher cutting range used in current study), i.e. 314 to 446 m/min is less than 8%. This difference is due to larger contact area at the lower speeds that heat distributes widely on the rake. This lower cutting speed range needs further investigation and more sophisticated and finer mesh over the entire rake face.

For higher speeds, i.e. 556 to 879 m/min, the difference is less than 6% (for both the paths). For the entire range of cutting speeds for both paths 1 and 2 the percentage difference between measured and simulated temperature is not more than 8%.

From the Figure it can be noted that difference in temperatures between FE-simulated and experimental results for six point matching for cutting speed 314m/min for path-1 (points 1 to 6) is less than 5.2% and for path-2 (points $1'$ to $6'$) is less than 4%. For cutting speed 556 m/min

(Figure) the difference between the experimental and FE temperature for path-1 is not greater than 5.16% and for path-2 it is less than 2.5%. The low difference between the experimental and FE temperature can be notd from Figure for the highest cutting speed, i.e. 879 m/min with 2.14% for path-1 and 1.8% for path-2.

Figure indicated the lowest temperatures on path-1 and path-2 as compared to the temperature at 314 m/min. This is because at higher speeds the contact length and also the sticking contact zone move towards the cutting edge and hence the temperatures near the contact length are higher but the temperatures distant from the contact area are low. Hence these higher temperatures confined in a narrow zone near the cutting edge and propagates flank wear and edge chipping. In general, FE-simulated temperatures and experimentally measured temperatures are in good agreement on two specified paths.

Figure shows heat partition (R_T) into multilayer-coated carbide cutting tool using the combined FE and experimental approach. It can be observed that as the cutting speed increases from 314 to 879 m/min, the heat partition (R_T) into the tool decreases to approximately 63%. At the cutting speed 314 m/min, the heat partition into the cutting tool is 12.9%. This heat partition is low when compared to the other studies, which used single-layer or two-layer coated tools. This low heat partition can be attributed to the coating layout, i.e. two-layer coatings on the rake and triple-layer coatings on the flank side and also to the restricted contact length.

For a cutting speed of 556m/min, the heat partition (R_T) into the cutting tool is 7.1% and at 879 m/min it is further reduced to 4.7%. Recently, Akbar et al. conducted a study to optimize the coatings with an aluminum content in the coating compound can reduce heat partition at 880 m/min to 3.5%. In the current study the topmost coating used is $Al_2 O_3$ on the tool rake face, which acts as a thermal barrier. Due to $Al_2 O_3$ decresing the thermal conductivity at higher temperatures, there was maximum resistance to heat flow to the tool and most the heat flows out from the system through the chip. At higher temperatues due to the lower thermal conductivity, the tool chip contact length decreases to the cutting edge. This is due to chip curel, i.e. thermal bimetallic effect (high difference in the thermal conductivity between workpiece and tool coating). The reduction in contact length forms a small contact area between the tool and the chip and hence causes low heat partition into the tool.

Furthermore, at higher temperature due to the presence of the $Al_2 O_3$ coating, aluminum from the coating oxidation during the cutting process is advantageous as it leads to a small amount of the workpiece material to stick onto the tool rake face. From Figure, it can be observed that iron concentration is found on the primary rake face, meaning that sticking is dominant in a very narrow zone (primary rake face) near the cutting edge. The small area and increases the sliding contact area and, hence, reduces the flow of heat into the cutting tool.

It can be concluded from the results that FE modeling of machining is challenging due to the complex nature of the process. So far considerable efforts have been made in this field, but FE modeling does not yet reflect the real machining process. This is due to the material properties, chip separation criteria, friction model, etc., which are important inputs for the reliable FE results. Therefore, simplified modeling techniques are needed, which use basic machining process parameters to model different machining characteristics.

In this study, transient thermal modeling of heat partition of the cutting wool was conducted. The methodology adopted for heat partition into the tool was a combination of experimental and FE simulation for complex coating layouts with restricted contact length/groove tools. It can be deduced from the current study that FE simulation output is based on the actual process parameters and true material properties. Thermal modeling of the machining process requires experimental data for the validation of the FE model. Using infrared thermometry for the validation of the FE model. Using infrared thermometry for the temperature measurement is entirely dependent on the emissivity of the material and this emissivity of the material changed with the change in several parameters. Hence, for reliable temperature measurement, emissivity should be evaluated.

Furthermore the hybrid methodology adopted in the current study reveals that by using different coating layouts and tool rake geometry, the heat partition (R_T) to the cutting tool can be reduced significantly. It can also be concluded that at high cutting speeds the contact zone shrinks to the cutting edge and hence rapid tool wear occurs, which is one of the main disadvantaged of HSM.

TUBE END-FORMING

Tube end-forming processes are commonly utilized for producing a wide variety of shapes and profiles, such as inversions, flares, expansions, reductions, beads and noses by means of single or multiple forming operations (Figure)

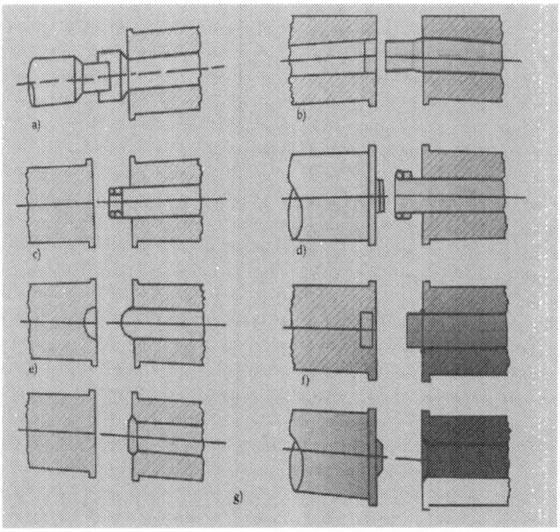

Figure 7.6: Tube end-forming processes: a) expansion; b) reduction; c) internal inversion; d) external inversion; e) nosing; f) compression beading; and g) flaring

Introduction

A historical review of the research on tube end forming can be organized into three different periods. The first period (1960-1990) draws from pioneering studies on axial loading to the external and internal inversion. During this period most attention was given to the identification of the main operative parameters, development of theoretical models to explain the mechanics of the processes, and the correlation between theory and experimentation.

The second period (1990-2000) maintained research focus on tube inversion and refined existing knowledge on th mechanics of deformation. The investigation on forming load and formability limits, by means of analytical methods, was progressively replaced by numerical simulation based on the finite element method in an effort to better understand the typical modes of deformation. In contrast to tube inversion, published research work in other tube end-forming processes during this period, was practically non-existent. At the end of the second period basic design rules were mainly derived from the accumulated experience of both manufacturers of tubular parts and suppliers of machine tools.

In what concerns the first objective, recent published work extended research to nosing, reduction, and expansion and flaring, and compression heading, among other tube end-forming processes.

The first technological validation studies ever performed with the objective of employing tube end-forming processes for shaping the end of thin-walled polymer tubes were recently published by Alves and Martins. Results show that manufacturers may extend their actual offer of polymer tubes to the production of custom and specific shapes at the tube ends. This is expected to open new market opportunities and stimulate innovations among tubular polymer products.

FEM in Tube End-forming

Finite element modeling supported by experimentation under laboratory controlled conditions is utilized for the understanding of the influence of major operative parameters, predicting the forming loads, identifying the modes of deformation, and setting-up the formability limits of the tube end-forming processes shown in Figure. The presentaition is restricted to a limited number of selected test cases made from aluminum and PVC.

In the first stage the load increases steeply as the tube is axially compressed against the surface of the die. A peak load of 18KN is obtained after which the tube starts to bend axially in order to match the contour of the die. Subsequently, the leading edge of the tube starts to turn around the die corner, as a consequence of progressive circumferential stretching, the load grows moderately from a lowered level. After the leading edge of the tube completes a $270°$ rotation the invert-forming process reaches steady-state conditions and, therefore, the value of the forming load P, becomes constant. Numerical and experimental curves compare well.

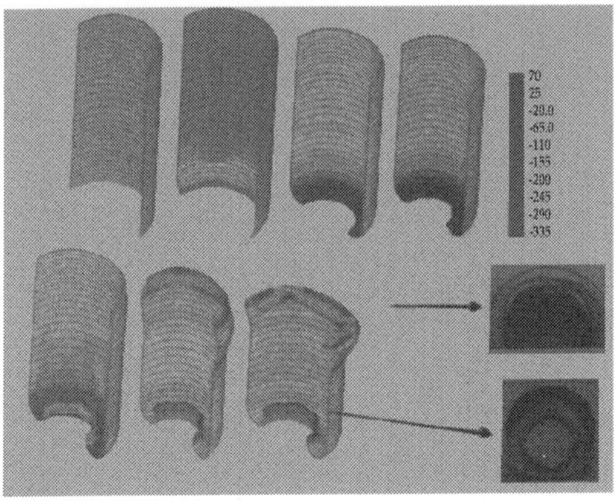

Figure 7.7: Predicted distribution of $\sigma_3\left(=\sigma_\theta\right)$(MPa) at intermediate stages of numerical simulation of the internal inversion of thin-walled tubes using large die radius values, ($\frac{r_d}{r_0} = 0.3, \frac{t_0}{r_0} = 0.05,$).Images show *significant* details of the final experimental tubular part

The predicted invert-forming load best matched the experimental results for values of the friction factor, m, in the range 0.02-0.09. In fact, as deformation progresses, it was found that lower values of friction should be employed for better reproduction of the experimental values. A possible explanation for this may be related to the built up pressure in the air trapped inside the unreformed portion of the tubular specimens due to the displacement of the upper flat die. It appears that the pressure acting on the lubricant placed at the entrance of the die-land can help to supply lubricant to the tool-workpiece contact interface, thus avoiding the breakdown of the lubricant film.

The role of lubrication was analyzed by utilizing a redesigned set of forming dies, which enabled variations in the lubrication regime. By allowing the air trapped inside the underformed portion of the tube to escape outside (through an air hole), it is no longer possible to guarantee a constant film thickness of the lubricant along the contact interface. Therefore, a significant portion of the lubricant film can break down or simply be removed during the silding of the tube and die can then contact each other and, as a consequence, an increase of the friction force enough to overtake the critical load is likely to occur at the end of the process.

The moment of load decline in the experimental test case performed with the modified tool was found to coincide with the formation of a fold at the top of the unreformed portion of the tube (near the upper flat die). Under these circumstances, it is possible to conclude that unacceptable local buckling of a successful inverse mode of deformation for $\frac{r_d}{r_0} = 0.278$ can easily take the place whenever the lubrication conditions become inadequate.

Finite element modeling of the external inversion of tubes was further assessed by comparing the predicted distribution of thickness strain in the cross-section of the tubular specimens with the experimentally determined values. The initial flat region of the graphic corresponds to the undeformed portion of the tube. The subsequent sharp increase in the thickness strain is related to the portion of the tube where it starts to bend in order to match the contour of the die. A peak value $\varepsilon_{th} \cong 0.069$ is predicted by finite element analysis after which the thickness strain gently decays as a consequence of progressive circumferential stretching ($\sigma_\theta > 0$). Negative values of thickness strain are encountered at the deformed portion of the tube undergoing near-rigid body motion. The final drop of thickness strain near the free curved end of the tube is a consequence of the initial non-steady state deformation mechanism in which the tubular specimen does not conform fully to the die. The grey-shaded contour placed on the top of the finite element mesh helps to visualize the differences between the experimental and the numerical prediction of the axial cross-section profile of an invert-formed specimen.

Experiments with internal inversion of tubes also revealed the existence of three different modes of deformation. For small values of the die radius formability is limited by the development of plastic instability modes (local bucking) whilst for larger values restrictions are normally set by excessive thickening of the tube wall and the occurrence of wrinkling at the unsupported curved wall of the tubes. As a consequence, material flow leading to successful inwardly inverted modes of deformation can only be achieved within a compact range of proves parameters.

Figure presents the theoretical and experimental load-displacement curve for a test case in which the internal inversion was accomplished successfully. Three phases may be distinguished; the axial compression stage, the transient invert-forming stage and the steady-state invert-forming stage.

In the first stage the load increases steeply as the tube is axially compressed against the surface of the die. A peak load of approximately 7KN is obtained after which the tube starts to bend axially in order to match the contour of the die. Subsequently (second stage), the leading edge of the tube starts to turn around the die corner and , as a consequence of compression along the circumferential direction è, the thickness of the tube wall progressively increases and the load grows moderately from a lowered level. After the leading edge of the tube completes a $270°$ rotation the invert-forming process reaches steady-state conditions and, therefore, the value of the forming load P, becomes constant.

During the experiments performed with larger values of the die radius it was possible to observe the occurrence of wrinkling at the free curved end of the tubes (refer to Figure). This failure is caused by high values of the tangential compressive stresses, σ_θ acting at the unsupported curved wall of the tubes. The overall agreement between the experimentally observed wrinkles and the correspondent finite element prediction is good and confirms the efficiency of the 3D numerical model in predicting material flow behavior under complex forming conditions.

The development of wrinkles and the thickening of the wall at the free curved end of the tubes increases the forming load necessary for the material to continue to flow inwardly beyond the critical instability load and causes subsequent development of local buckling at the undedormed portion of the tube.

To conclude this section on invert forming of thin-walled tubes it is important to understand that numerical and experimental simulation performed at different lubrication regimes revealed that successful inversion modes could be easily replaced by non-successful modes and vice-versa, which provides evidence for the importance of lubrication in the overall formability of the process. The existence of an adequate lubricant film along the contact interface between the tube and the die is essential for the overall success of the invert-forming of double-walled tubular parts.

The investigation in the field also revealed the main operative parameters that govern the inversion of tubes using a die; (1) the ratio between the fillet radius of the die and the outer radius of the tube, (2) the ratio between the wall thickness and the outer radius of the tube, and (3) the height h of the straight part of the die, in the case of internal inversion. That last parameter influences the alignment between the undeformed parts of the tube and dies during the internal inversion.

Expansion and reduction of tubes is the result of three different mechanisms (Figure): bending/unbending, stretching/compression along the circumferential direction, è (for expansion/reduction of tubes), and friction. Bending takes place at point B, where the tube first contacts the punch (or die), while unbending takes places at point D, where the tube leaves contact with the punch (or die). Point E fixes the beginning of the free deformed region of the tube, which spans from E to the leading edge F. The free deformed region of the tube undergoes near-rigid body motion. Stretching/compression develops while the tube gradually deforms over the punch (or die; refer to the cylindrical surface adjacent to point B).

The experimental work on tube expansion confirmed the existence of three different modes of deformation. For large ratio of and high value of the angle á , formability is limited by the occurrence of plastic instability while for larger lengths of expansion (á ≪) restrictions are normally set by ductile damage (fracture) in the regions that are highly stretched in the circumferential direction. The experimental work on tube reduction also revealed the existence of three different modes of deformation. For small values of and small lengths of reduction formability is mainly limited by the development of plastic instability modes whilst for larger lengths reduction restrictions are normally set by excessive thickening of the tube wall and by the occurrence of wrinkling at the conical wall of the tubes.

Modeling the occurrence of cracking in tube expansion was accomplished by means of a 3D finite element analysis that made use of physical separation of adjacent elements at the nodal points where ductile damage, according to the Cockcroft-Latham normalized criterion, reached critical values. The agreement between experimental and numerical predicted shapes resulting from finite element analysis is good and the theoretical distribution of damage, obtained shortly after the beginning of fracture, successfully estimates its peak value to occur at the tip of the crack located at the upper end of the tube as seen in the experiments.

In the case of tube reduction formability limits result from a wrinkling collapse mechanism. In fact, besides the commonly acknowledged effect of the compressive circumferential stress, $\sigma_\theta = \sigma_3$, acting at the unsupported conical region of the tubes, it is also possible to conclude that the excessive thickening of the upper tube wall will create extra resistance to deformation that will be responsible for building up a peak forming load within the range of instability ($P \cong P_{cr}$). This is the reason why tubular specimens usually buckle shortly after the occurrence of the first wrinkles.

The essence of the proposed wrinkling collapse mechanism is corroborated by experimental and finite element predicted shapes that are depicted in Figure. Again, an excellent agreement is found between numerical and experimental result.

Verification

Conventional processing technologies for tubular polymer parts require large batch sizes (mass production) because the energy costs and the capital investment in equipment and tooling (e.g. machine-tools, mounds, dies, jigs, and fixtures) are very high. This makes conventional processing techniques unsuitable for extending the present offer of secondary operations that can be performed after the main production stage. However, the reader might ask what would occur if end-forming operations currently utilized in metal tubes, such as expansion, reduction, inversion, flaring, flanging, and tapering, were applied to polymer tubes.

If tests prove that sound shapes can be successfully produced in polymers by means of tube end-forming operations, it appears obvious to embrace and extend some of the existing technologies to polymers.

Nevertheless, end forming of polymer tubes represents a technological challenge for the current state-of-the art polymer processing because the mechanisms of cold plastic deformation of polymers (which are quite different from those of metals) need to be fully understood to be able to establish process formability limits in terms of the major variables. Only then will it be possible to provide engineers with the basic design rules for forming the required tubular end shapes.

The work on external inversion of tubes showed that for small values of restrictions are normally set by fracture in the regions that are highly strained due to extensive material flow.

When the external inversion is successfully accomplished the experimental load-displacement curve allows the identification of three different forming stages: the axial compression stage, the transient invert-forming stage, and the stead-state invert-forming stage.

In the axial compression stage, the load increases steeply as the tube is axially compressed against the surface of the die. A peak load of approximately 8KN is obtained after which the tube starts to bend axially in order to match the contour of the die.

Subsequently (second stage), the leading edge of the tube starts to turn around the die corner and, as a consequence of progressive circumferential stretching, the load grows moderately from a lowered level. After the leading edge of the tube completes a 270°C rotation, the invert-forming process reaches steady-state conditions and, therefore, the value of the forming load becomes constant and approximately equal to 6KN.

The agreement between experimental and finite element estimates of the load-displacement curve is good. Major differences are found to occur in the vicinity of the peak load at the first stage of the process.

The load-displacement curve of mode 1 is different from that of mode 3 because the axial compression stage (i.e. the first stage of deformation) experiences a steep increase of the forming load coincides with the formation of a polygonal-shaped fold at the top of the undeformed region of the tube located near the upper flat die.

Under these circumstances, it is possible to set up the lower limit of the curling ratio of PVC tubes within the range 0.1 to 0.15. The upper limit of the curling ratio is limited by fracture (refer to mode 3 in Figure) and the corresponding mechanism results from the accumulation of ductile damage.

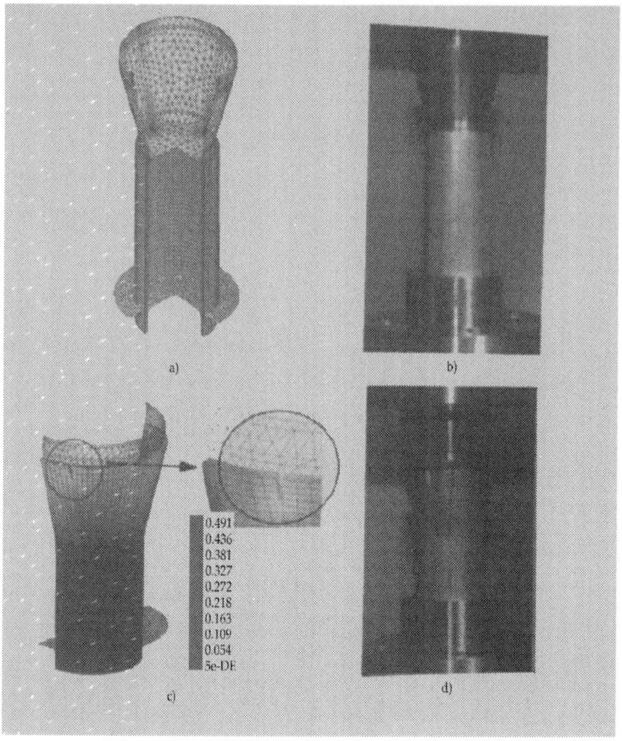

Figure7.8: Expansion of tubes ($t_0 = 2\ mm, \frac{r_p}{r_0} = 1.76, \text{ and } \alpha = 15°$). a) 3D finite element model; b) experimental set-up at the beginning of deformation; c) theoretical distribution of ductile damage (normalized Cockcroft-Latham) after 45 mm of punch displacement; and d) experimental part after an amount of displacement equal to c)

The work on internal inversion of tubes also revealed the existence of three different modes of deformation. For small values of the curling ratio, formability is limited by the development of plastic instability (local buckling) whilst for larger values restrictions are normally set by excessive thickening of the tube wall followed by plastic instability.

The work on expansion and reduction of PVC tubes also revealed the existence of three different modes of deformation. In the case of expansion, the first example shows a fracture, the second and third examples illustrate forming defects resulting from local buckling, and the fourth example presents a successful component. In the case of reduction, the first and third examples show forming defects resulting from local buckling and the second example presents a successful component.

In tube expansion PVC is submitted to strain hardening and damage. These two competing mechanisms control the feasibility of the process as well as the final mechanical properties of the tubular formed parts. The distribution of damage can be obtained from a decoupled finite element analysis based on the utilization of the normalized Cockcroft-Latham ductile fracture criterion.

The computed distribution of damage qualitatively corroborates the experimental location of the onset of cracking. The highest value is found to occur at the upper edge of the tube and its magnitude provides a good estimate of the critical local of damage in PVC.

PVC also experiences a certain amount of crazing during tube expansion. Crazes are crack-like defects (but are not true cracks) that are characterizedbyregionsofhighlyplasticallydefomedmaterialinterspersed with voids and are expected to be triggered when the normalized ratio of the stress-bias criterion $\frac{\sigma_1}{\sigma_0} > 1$, However, the phenomenon is limited to the leading edge of the tubes because the compression pressure imposed by the punch on the conical wall contributed to lower the overall level of hydrostatic stresses.

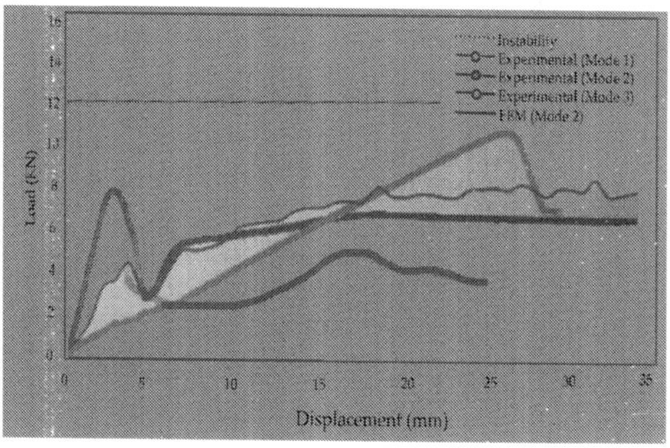

Figure 7.9: Load-displacement curve for the external inversion of PVC tubes using a die

When the same test case is performed under conditions of poor lubrication, it is experimentally observed that the tube preferentially develops local buckling. No catastrophic failure due to cracking is found to occur. The reason for this is that after a certain amount of free expansion, the leading edge of the tube stops flowing along the conical surface of the punch and, as a result of the ongoing axial compression, the tube starts forming an instability wave near the bottom-die. The change in material flow cancels the building up of damage at the upper edge of the tube and promotes its rapid accumulation at the outside regions of the instability wave.

The amount of crazing until the tube stops flowing along the conical surface of the punch is also limited. In fact, the normalized stress-bias criterion stays $\frac{\sigma_1}{\sigma_0} < 1$ along the conical wall of the tube apart from a small region, located at the upper leading edge, which overtakes the limiting threshold value $\frac{\sigma_1}{\sigma_0} = 1$ at the early stages of the tube is lighter than the original gray color of the tube at the beginning of tube expansion.

The subsequent rise in tensile hydrostatic stresses at the instability tube wave near the bottom-die, leads to a significant propagation of crazes as can be deduced from the changes in color that take place in that region. The agreement between experimental observation and finite element distribution of the normalized stress-bias criterion is very good and suggests that it can be successfully utilized for predicting the location inside PVC tubular parts where crazing is likely to occur.

Design rules and process parameters for the end forming of tubes using a die must be properly established so that material deformation does not approach formability limits due to ductile fracture, local buckling, and wrinkling. Tube wall thickness and lubrication play a key role in obtaining a successful mode of deformation.

Admissible values of the major process parameters are usually bounded by values that give rise to buckling and values that stimulate the occurrence of cracking or wrinkling. The utilization of ductile fracture criteria in conjunction with numerical modeling by finite elements can help predict the onset of fracture.

The good agreement found between the experimentally measured and the computer-predicted shapes, formability limits, thickness variations, and load-displacement evolutions allow us to conclude on the adequcacy of the extended finite element flow formulation to model cold plastic deformation of metallic polymer and bi-material tubular parts.

ROLLING PROCESS

Rolling is a metal-forming process, in which raw material is shaped by passing it between two counter-rotating cylinders called rolls. In this process the desired shape is obtained by plastic deformation. Depending on the desired shape, the rolling is classified as plate rolling, strip rolling, slab rolling, rod rolling, thread rolling, etc. The aim of modeling the rolling process is to investigate the rolling force, roll torque, stress, strain, strain rates, and temperature as a function of process and material parameters.

Introduction

Starting from the work of von Karman, the rolling processes have been analyzed by a number of researchers. Some of the traditional modeling methods are slab method, slib line method, and upper bound method. The slab method is one of the simplest methods that can be applied for analyzing the rolling process. In this method, a slab of infinitesimal thickness is selected perpendicular to the flow direction at an arbitrary point in the deformation zone. Force and momentum balance for the slab results in differential equations, which are solved either by closed-form analytical methods or numerical methods. The constants of integration are evaluated using appropriate boundary conditions.

The slab method uses several simplifying assumptions. For example, in a symmetric flat rolling process, it is assumed that longitudinal direct stress is constant across the thickness and it is one of the principle stresses. Non-uniformity in the thickness direction has been considered in slab method modeling of asymmetric rolling by making the assumption that the longitudinal stress is linearly distributed across the thickness direction. Due to simplifying assumptions, the slab method model is not useful for carrying out highly accurate and detailed analysis of rolling processes.

The use of slip line method to solve plane strain problems at each point in the plane of plastic flow considers a pair of orthogonal curves along which the shear stress is at a maximum. The pair of orthogonal curves are called slip lines as the material slips along these lines. Equilibrium equations, incompressibility relation, yield criterion, stress-strain relations and boundary conditions determine the stress distribution during the rolling process. The slip-line field solution for cold rolling were suggested by Firbank and Lancaster. Nepershin modeled a steady-state plane strain slab and strip hot rolling problem using the silp line method for the rigid-plastic material. The slip line method cannot be used universally because a) The method is applicable only for plane strain problems; b) the slip line construction is a difficult task; c) It is not easy to incorporate strain hardening, strain-rate, and temperature effects in the slip line method; d) It is not possible to incorporate elastic effects in this model.

In the upper bound method, a kinematically admissible velocity field, which satisfies the continuity equation and velocity boundary conditions, is assumed. The total power obtained based on the assumed velocity filed provides an upper bound on the actual power. Although the method provides only limited and approximate information about the rolling process, it has been researched by a number of author.

The finite element method has established itself as a robust and efficient method of solving the governing differential equations that have been applied to a number of engineering problems. It has been used by a number of researchers to model rolling problems. The FEM can easily incorporate non-homogeneity of deformation, process-dependent material properties, and different friction models.

FEM in Rolling Process

The FEM is a numerical method of solving differential equations. In this method, the domain is discretized into a number of small elements and the primary dependent variables of the differential equations are approximated by suitable functions over an element. The form of the function is assumed beforehand, but certain parameters are taken as unknowns. These unknown parameters are obtained in process of minimizing the error. Different error measures can be chosen, which provides a variety of FEM formulations.

In an element, certain points are designed as nodes. Often, the nodal values of primary variables replace the approximating parameters of the approximation function. Thus, in the FEM unknown nodal values are determined. In addition to unknown nodal values, some other parameters associated with the element may also be obtained.

As a result of finite element procedure, algebraic equations are obtained instead of the terms containing differentiation with respect to spatial coordinates. The elemental equations are assembled to obtain the equations for the global system. The purpose of the assembly is to eliminate some internal constants and obtain as many equations as there are unknown nodal or parameter values.

The most common method of obtaining the elemental equations in the FEM is the Galerkin method, which is briefly explained here. Consider the differential equation $D(u) = 0$, where u denotes a vector of primary (dependent) variables. Inside an element of volume V, u can be approximated by $\bar{u}$. Clearly, $D(\bar{u}) \neq 0$, unless the approximation is exact. Writing $D(\bar{u}) \neq R$, the ideal objective of any numerical method will be to make residual R zero. In the weighted residual method, we multiply residual R by a weight function w and make the volume intergral of wR equal to zero, i.e.

$$\int_V wRdV = \int_v wD(\bar{U}) = 0$$

In this equation, the order of derivative of w is 0, whilst the highest order of derivation of $\bar{u}$ is equal to the order of the differential equation. We integrate equation in parts, in order to reduce the order of derivative in overall integrand. The resulting expression, which contains boundary terms as well, is called the weak form of equation. The approximating function of the primary variable is constructed in such a manner so that the highest derivative of the primary variable in the weak form should be non-zero. Additionally, at the interface of two elements, the approximating functions should have the continuity of at least m-1, where m is the highest order derivative of $\bar{u}$ in the integrand of the weak form.

For an n-noded element, the primary variable $\bar{u}$ may be expressed as:

$$\bar{u} = \sum_{i=1}^{n} N_i u_i$$

Where u_i is the nodal value of primary variables, a_i is the additional parameter of the element, and N_i is the shape functions. The commonly used elements in 2D are triangular and quadrilateral. In 3D, tetrahedral or hexahedron elements are used. For developing the equations, the elements are transformed from physical to natural coordinates. When the geometry is approximately by the same degree, the formulation is called iso-parametric formulation. When a higher degree of approximation is used for geometry compared to primary variables, the formulation is called a super-parametric formulation. When a higher degree of approximation is used for primary variables compared to geometry, the formulation is called a sub-parametric formulation.

Verification

In hot rolling, an equation showing the dependency of flow stress on strain, strain rate, and temperature is used. A number of equations have been proposed. Here, some equations are described. The equations have been written in their 1D form. The general form can be obtained by replacing 1D stress, strain, and strain rates by the corresponding equivalent variable. One relation is as follows:

$$\sigma = K\varepsilon^n \dot{\varepsilon}^m \exp\left(\frac{\beta}{T}\right)$$

Where K, n, m and β are material constants. In the general form σ will be replaced by σ_{eq}, ε by ε_{eq} and $\dot{\varepsilon}$ by $\dot{\varepsilon}_{eq}$. The hyperbolic law in an Arrhenius type of equation is given by:

$$Z = \dot{\varepsilon} \exp\left(\frac{Q}{RT}\right) = A\left[\sinh\left(\alpha\sigma_p\right)\right]^n$$

Where σ_p is the peak stress, Z is the Zener-Holloman parameter, Q is the activation energy, R is the universal gas constant, and A, á and n are material constants. The equation accounting for softening is given by:

$$\sigma = \sigma_p\left[1-\exp(-C\varepsilon)\right]^m - \left(\sigma_p - \sigma_s\right)\left[1-\exp\left\{-k\left(\frac{\varepsilon-\alpha\varepsilon_p}{\varepsilon_p}\right)^{m'}\right\}\right]$$

Where ε_p is the peak strain, σ_s is the steady-state stress. The constants appearing in the above equation are further dependent on temperature and strain rate. These have to be empirically determined for each material.

Another relation for flow stress is given as:

$$\sigma = \sqrt{3}\exp\left(\frac{\beta}{T}\right)\left(\varepsilon + \varepsilon_0\right)^n\left(K_0 + K_1\left(\sqrt{3}\dot{\varepsilon}\right)^m\right)$$

Where the constants appearing in the equation have to be determined experimentally. Duan and Sheppard have compared three constitution equations in the FEM modeling of hot rolling. One equation is Norton-Hoff law, Another equation is a hyperbolic sine function given by:

Where A, σ and n are constants. The third equation is the Hensel and Spittel equation given by:

$$\sigma = A\exp\left(m_1 T\right)\varepsilon^{m_2}\dot{\varepsilon}^{m_3}\exp\left(\frac{m_4}{\varepsilon}\right)$$

Where A, m_1, m_2, m_3 and m_4 are material constants. During steady-state conditions, the measure roll force was 0.879 t/mm. Predicted forces are 0.865, 0.900, and 0.755 t/mm from the Norton-Hoff law is very close to experimental result. The hyperbolic sine function overpredicts and Hensel and Spitel equation under-predicts the roll pressure. There is no significant difference in the temperature prediction in three cases.

The trend is similar to roll force prediction. Most of the energy of plastic deformation is converted to heat. Therefore, it is obvious that the higher the roll force, the higher the rise in temperature. However, as other thermal properties and heat transfer coefficients are the same in the FEM computation, the effect of variation of roll force is negligible on temperature prediction.

Figure shows the variation of computed temperatures at the center during the second rolling pass. It is observed that all three material models predict almost same temperature. It is observed that under steady-state conditions the hyperbolic sine function predicts less than the other two models. Actually, for the second rolling pass, even the roll force prediction was on the lower side for the hyperbolic sine function model.

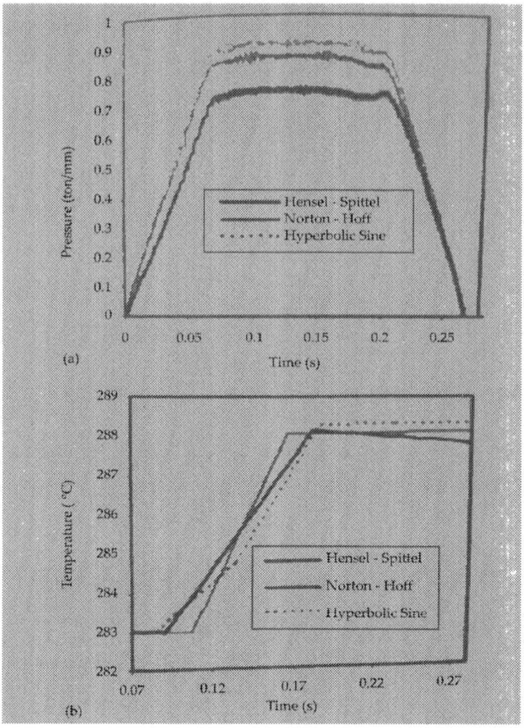

Figure 7.10: Computed histories of first rolling pass: a) roll pressure; b) temperature at the center point

Instead of using the material model in the form of equations, neural networks can be used to model material behavior. Neural networks are suitable for any online mapping; however, they require sufficient data and skill in modeling.

In hot rolling, heat transfer and mechanical analysis needs to be carried out simultaneously. In general, the solution is sensitive to boundary conditions and material behavior, any developed finite element model must be validated from the experiments. A rigorous finite element analysis requires enormous computational time, which can be reduced by carrying out the neural network-assisted finite element analysis. In addition, a neural network can be trained using the results from the finite element analysis. The trained network can be used for online prediction. Microstructural modeling and defect analysis in rolling is one area where the finite element analysis can play a major role.

CHAPTER 8

FINITE ELEMENT METHOD IN AEROSPACE ENGINEERING

INTRODUCTION

Structures are often analyzed using complex finite element analysis methods. These tools have evolved over the past decades (since early 1960's) to be the basis of most structural design tasks. A candidate structure is analyzed subject to the predicted loads and the finite element program predicts deflections, stresses, strains, and even buckling of the many elements. The designed can then resize components to reduce weight or prevent failure. In recent years, structural optimization has been combined with finite element analysis to determine component gauges that may minimize weight subject to a number of constraints. Such tools are becoming very useful and there are many examples of substantial weight reduction using these methods. Surprisingly, however, it appears that modern methods do not do a better job of predicting failure of the resulting designs, as shown by the figure below, constructed from recent Air Force data. The data compares static test failures of wing, fuselage, vertical tail, horizontal tail, landing gear and other components for New

Aircraft (designed with finite element methods) and Older aircraft (designed without FEM).

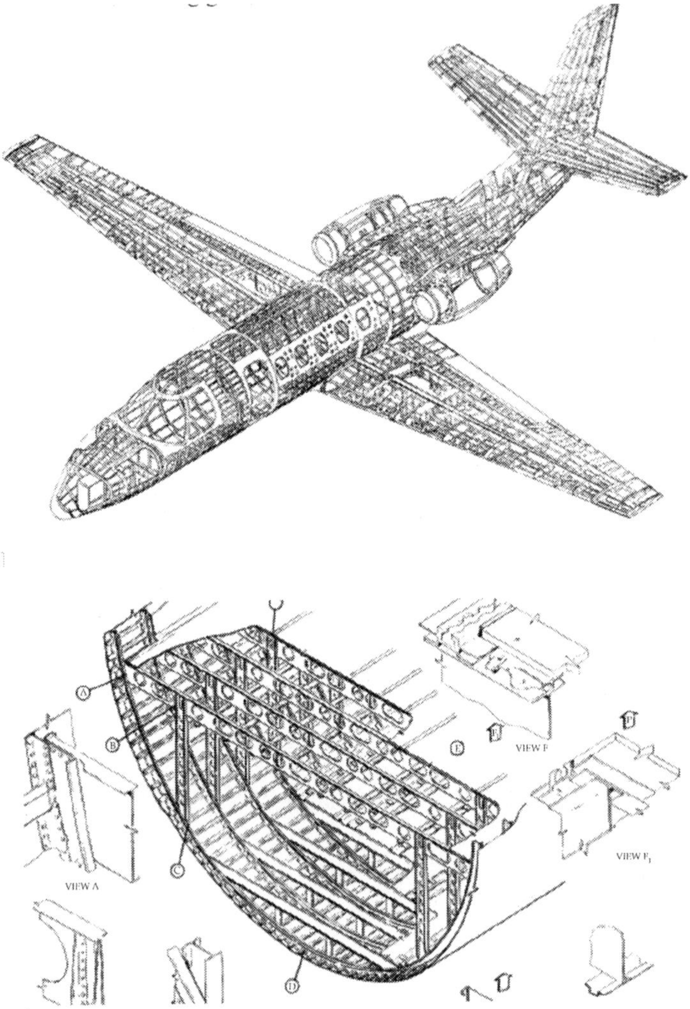

Figure 8.1: Structure of aircraft.

Structural members shown above may be variously modeled as beams, thin plates, membranes, shells, etc. In some cases, it may be necessary to perform a full 3-D stress analysis (an elasticity type analysis as opposed to an approximate one like plate theory).

The webs in wing ribs and floor beams, and wing and fuselage skins are typically thin members that might be considered as being in a state of plane stress.

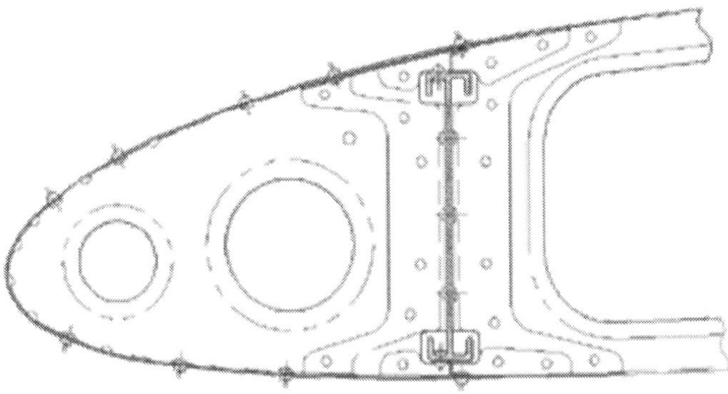

Figure 8.3: Plane structure.

Plane stress describes a three-dimensional geometry wherein the non-zero stresses all occur in a single plane. For a thin plate, web or skin lying in the x-y plane, the only non-zero stress components are $\sigma_{xx}, \sigma_{yy}, \sigma_{xy}$.

We now consider the development of a plane stress finite element. Suppose we have a geometry like that shown below where the thickness is small compared to the other two dimensions:

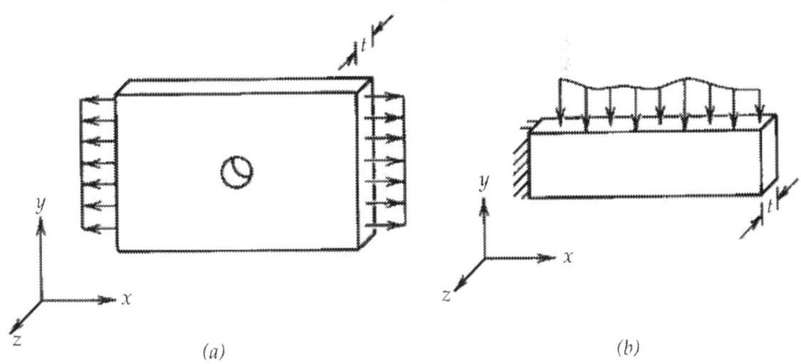

Figure 8.3: Stress components.

The only major stress components are $\sigma_{xx}, \sigma_{yy}, \sigma_{xy}$. Consider another example of a thin bracket in tension and the corresponding finite element mesh. The boundary at the hole is welded and therefore fixed. A static tensile load of 5,500 lbs is applied to one end. The dimensions are 25" x 10" x 0.5" thick.

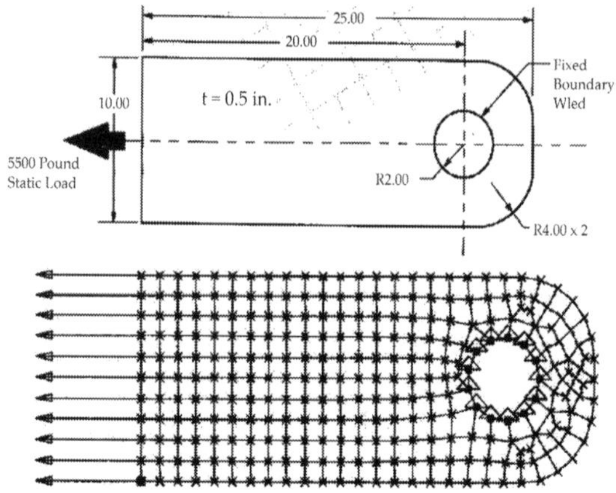

Figure 8.4: Bracket geometry and FEM mesh, loading and BC.

Notice that the geometry has been divided up into a number of rectangular regions (elements) - these are called Quad elements.

We could also use triangular elements. We will demonstrate the development of the stiffness matrix and load vector for a triangular element as shown below. We assume that plane stress occurs in the x-y plane and define displacement components and v(x, y). We define nodal displacements at the three corners as:

Node 1: u_1 and v_1

Node 2: u_2 and v_2

Node 3: u_3 and v_3

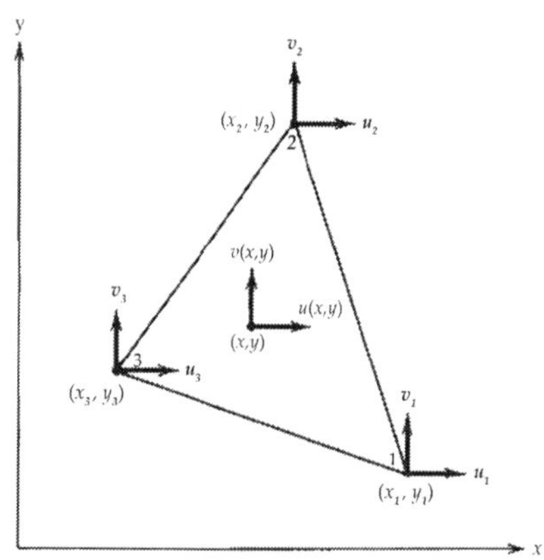

Figure 8.5: Constant strain triangle (CST) and its six nodal degrees of freedom.

The process of developing the equilibrium equations for a given element requires that we utilize an energy or variational principle; for example, the principle of minimum potential energy given by:

$$\delta(U+V)=0$$

where U is the internal strain energy and V is the external potential energy.

For a plane stress state, the internal (potential) energy is given by

$$U=\tfrac{1}{2}\int_V \{\sigma\}^T\{\varepsilon\}dV$$

Where

$$\{\sigma\}=\begin{Bmatrix}\sigma_{xx}\\ \sigma_{yy}\\ \sigma_{xy}\end{Bmatrix} \quad and \quad \{\varepsilon\}=\begin{Bmatrix}\varepsilon_{xx}\\ \varepsilon_{yy}\\ \varepsilon_{xy}\end{Bmatrix}$$

Assuming a linear elastic material, the constitutive equation may be written as

$$\{\sigma\}=[D]\{\varepsilon\}$$

where [D] for plane stress is given by:

$$[D]=\frac{E}{1-\nu^2}\begin{bmatrix}1 & \nu & 0\\ \nu & 1 & 0\\ 0 & 0 & (1-\nu)/2\end{bmatrix}$$

Note that [D] is symmetric.

The infinitesimal strains are given:

$$\varepsilon_{xx}=\frac{\partial u}{\partial x}, \qquad \varepsilon_{yy}=\frac{\partial v}{\partial y}, \qquad \varepsilon_{xy}=\frac{\partial u}{\partial y}+\frac{\partial v}{\partial x}$$

where u(x, y) and v(x, y) are the displacement fields.

the strain energy is

$$U = \frac{1}{2}\int_V \{\varepsilon\}^T [D]\{\varepsilon\} dV$$

where use has been made of the symmetry of [D]. U is now in terms of the displacement fields u(x, y) and v(x, y).

The external potential V can be evaluated once the external tractions and body forces are specified. In general, V will have the form of

$$V = -\int_S p_x u + p_y v)ds - \int_V (F_x u + F_y v)dV$$

where p_x and p_y are boundary tractions, S is the element boundary surface, F_x and F_y are body forces. Note that V is in terms of displacements u(x, y) and v(x, y).

We now have U and V in terms of u(x, y) and v(x, y); and the equilibrium equations for are an element are defined by the energy principle. However, we can't apply it just yet. Why? Because u(x, y) and v(x, y) must be in terms of discrete variables (nodal displacements) but u(x, y) and v(x, y) are continuous functions.

Assemblage of Elements

A single element by itself is useless. We must determine the equilibrium equations for an assemblage of elements that comprise the entire structures.

Consider the following structure (only a few elements are taken to simplify the discussion) with a uniformly pressure p on the right boundary and fixed on the left boundary (assume a constant thickness t).

We number the structural nodes from 1 to 12 as shown. We also number the elements from 1 to 12 as shown (in any order).

For each global node of the structure, we can specify the (x,y) coordinates: x_i, y_i, i=1, 2, …, 12.

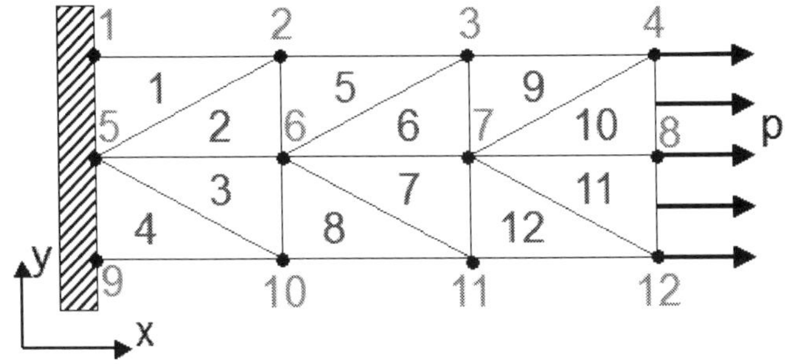

Each node of the structure will have two degrees of freedom (dof). We label these structural (global) degrees of freedom in order as shown to the right. Note that the structural nodal displacements are written without the superscript "e." The nodal displacement vector is written as and is (24x1) for this problem.

We note that the left side is fixed (nodes 1, 5 and 9). Hence, displacement boundary conditions will require that

$$q_1 = q_2 = q_9 = q_{10} = q_{17} = q_{18} = 0$$

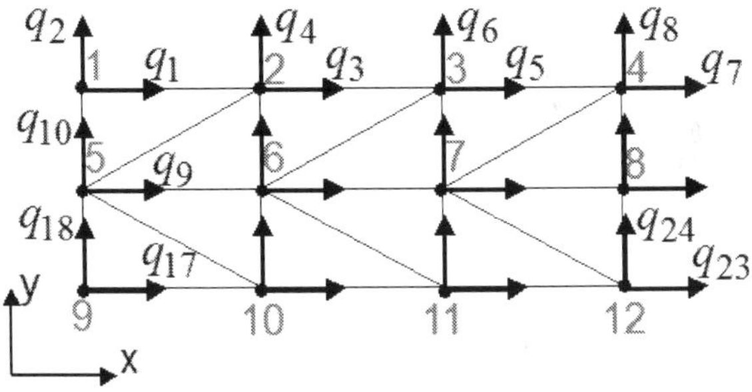

Note: we do not have to number the dof consistently and in sequence with the structural nodes. However, this makes the bookkeeping much, much simpler!

For each element, we can construct a table called the element connectivity that specifies which structural (global) nodes are connected by an element. Hence, for the problem above, we have the following element connectivity table:

Element No.	Element Node 1	Element Node 2	Element Node 3
1	1	5	2
2	5	6	2
3	5	10	6
4	5	9	10
5	2	6	3
6	6	7	3
7	6	11	7
8	6	10	11
9	3	7	4
10	7	8	4
11	7	12	8
12	7	11	12

Note that for the CST, element nodes MUST be given as CCW. Element node 1 can be attached with any global node of the element.

Note that if we are careful in numbering the nodes and choosing the element connectivity in a "systematic" manner, there will be a pattern to the element connectivity table (see above). An automatic mesh generator, like the one in FEMAP, tries to follow this pattern.

Note that the global node numbers for the structure are somewhat arbitrary, i.e., we could number them in any order. However, it will turn out that there are optimum ways to number nodes (for a given structure and mesh) in order to reduce the bandwidth of the structural stiffness matrix [K] - this saves time solving the equations. For the mesh above, it would be optimum to number downward and left-to-right, as opposed to left-to-right and downward. We'll discuss that later. Likewise, the element numbering is arbitrary, but again there may be optimum approaches. An automatic mesh generator tries to do the numbering in an optimum fashion.

Note that for this structure, we have 12 global nodes. There are 2 degrees of freedom (dof) at each node (u and v). Hence, the structure has 24 dof and the structural stiffness matrix [K] will be (24x24). The structural equilibrium equations can be written as:

$$\underset{(24x24)}{[K]}\ \underset{(24x1)}{\{q\}} = \underset{(24x1)}{\{Q\} + \{S\}}$$

where [K]=structural stiffness matrix,

{Q}=structural forces matrix (due to applied tractions and body forces)

{S}=structure reaction forces due to boundary conditions

Let's see how each element contributes to global matrices. Take element 1 to start with. Note that we can use sub-matrix notation to divide the element matrices as following. Use a superscript of 1 on the k terms to indicate element 1.

$$\underset{(6x6)}{[k^1]} = \begin{bmatrix} \underset{(2x2)}{k_{11}^1} & k_{12}^1 & k_{13}^1 \\ k_{21}^1 & k_{22}^1 & k_{23}^1 \\ k_{31}^1 & k_{32}^1 & k_{33}^1 \end{bmatrix}, \quad \{F^1\} = \begin{Bmatrix} \underset{(2x1)}{F_1^1} \\ F_2^1 \\ F_3^1 \end{Bmatrix}$$

We now look at element 1 and note that element node numbers 1, 2, 3 correspond to global node numbers 1, 5, 2 (from the drawing of the mesh, or from the element connectivity table). We can indicate this information on the stiffness and force matrices as follows:

$$\underset{(6x6)}{[k^1]} = \begin{bmatrix} \underset{(2x2)}{k_{11}^1} & k_{12}^1 & k_{13}^1 \\ k_{21}^1 & k_{22}^1 & k_{23}^1 \\ k_{31}^1 & k_{32}^1 & k_{33}^1 \end{bmatrix}, \quad \underset{(6x1)}{\{F^1\}} = \begin{Bmatrix} \underset{(2x1)}{F_1^1} \\ F_2^1 \\ F_3^1 \end{Bmatrix}$$

Hence, we see that element 1 contributes stiffness and forces to global nodes 1, 5 and 2. Placing these contributions into the global stiffness matrix gives:

Element 1 only

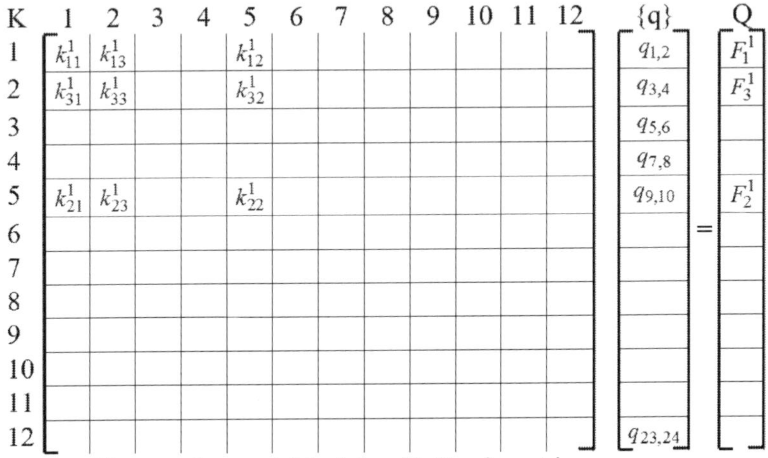

** remember, each block is a (2x2) sub-matrix

Now take element 7.

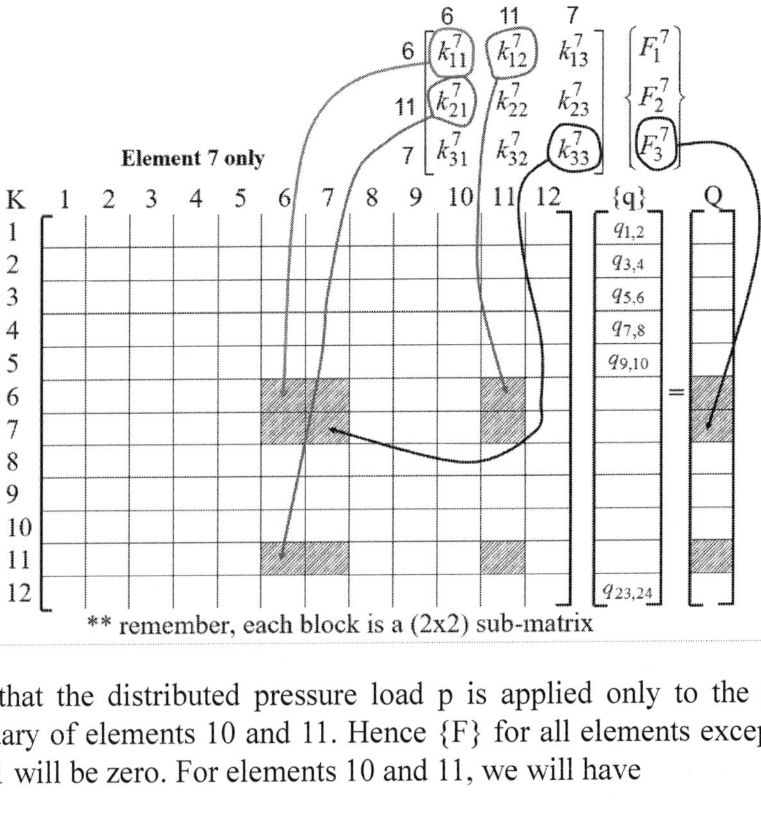

** remember, each block is a (2x2) sub-matrix

Note that the distributed pressure load p is applied only to the right boundary of elements 10 and 11. Hence {F} for all elements except 10 and 11 will be zero. For elements 10 and 11, we will have

$$\{F^{10}\} = \left\{ \begin{array}{c} 0 \\ 0 \\ \hdashline \frac{1}{2}ptL_{4-8} \\ 0 \\ \hdashline \frac{1}{2}ptL_{4-8} \\ 0 \end{array} \right\} \begin{array}{c} 7 \\ \\ 8 \\ \\ 4 \\ \end{array} \qquad \{F^{11}\} = \left\{ \begin{array}{c} 0 \\ 0 \\ \hdashline \frac{1}{2}ptL_{8-12} \\ 0 \\ \hdashline \frac{1}{2}ptL_{8-12} \\ 0 \end{array} \right\} \begin{array}{c} 7 \\ \\ 12 \\ \\ 8 \\ \end{array}$$

where L_{4-8} is the length between global nodes 4 and 8, etc.

If we assemble all element stiffness matrices [k] and forces matrices {F} to the global equilibrium equations, we have the following result:

Structural Equations of Equilibrium

K	1	2	3	4	5	6	7	8	9	10	11	12	{q}		Q
1	X	X			X								$q_{1,2}$		
2	X	X	X		X	X							$q_{3,4}$		
3		X	X	X		X	X						$q_{5,6}$		
4			X	X			X	X					$q_{7,8}$		X
5	X	X			X	X			X	X			$q_{9,10}$		
6		X	X		X	X	X			X	X			=	
7			X	X		X	X	X			X	X			
8				X			X	X				X			X
9					X				X	X					
10					X	X			X	X	X				
11						X	X			X	X	X			
12							X	X			X	X	$q_{23,24}$		X

X means that one or more elements have contributed here
** remember, each block is a (2x2) sub-matrix

Note that [K] is symmetric; also it is banded (semi-bandwidth=12).

In the previous page, each X means that one or more elements have contributed to that (2x2) sub-matrix. For example, we note that node 2 will have stiffness from elements 1, 2 and 5. Hence, the 2,2 position of the global stiffness matrix will be equal to (note: you have to refer to the element connectivity to see which element node for each element corresponds to global node 2):

$[K_{22}]=[k_{33}^{1}]+[k_{33}^{2}]+[k_{11}^{5}]$

The global node 2-6 coupling term $[K_{26}]$will have contributions from elements 2 and 5 since only these elements share the boundary between nodes 2 and 6:

$$[K_{26}]=[k_{32}^{2}]+[k_{12}^{5}]$$

Global node 6 will have stiffness contributions from elements 2, 3, 5, 6, 7, and 8:

$$[K_{66}]=[k_{22}^{2}]+[k_{33}^{3}]+[k_{22}^{5}]+[k_{11}^{6}]+[k_{11}^{7}]+[k_{11}^{8}].$$

Question? What happened to the reactions {S} for each element? Why don't they show up in the structural stiffness matrix?

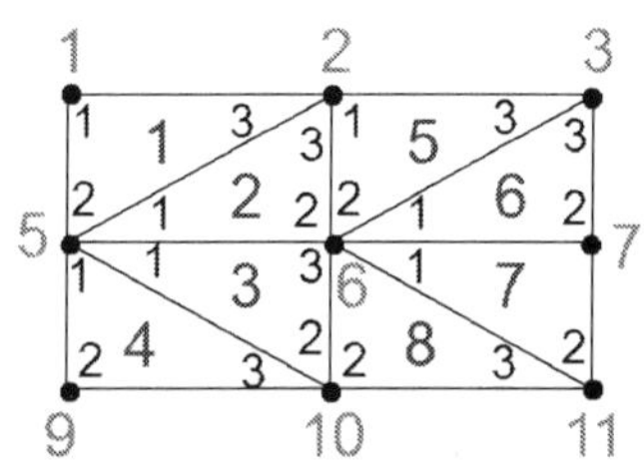

Simple. It is equilibrium. Recall that when we make a free-body, in this case take a single finite element as the free-body, we will have equal and opposite reactions where the cut is made though the body. Consider elements 1 and 2 below:

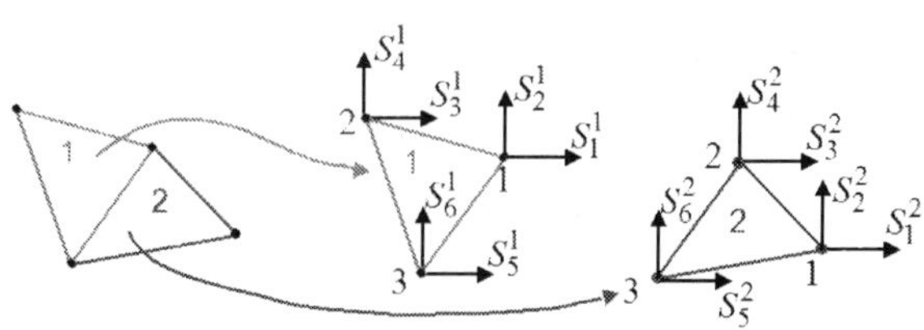

At the boundary between elements 1 and 2, the reactions are equal and opposite. Hence, we add them up we have: $S_1^1+S_3^2=0$, $S_2^1+S_4^2=0$, $S_5^1+S_5^2=0$, and $S_6^1+S_6^2=0$. Hence, all the reactions between elements sum to zero and do not have to be put into the structural equilibrium equations.

OK, but what about the boundary where there are supports? What happens to the reactions there? For example, the cantilever plate example above:

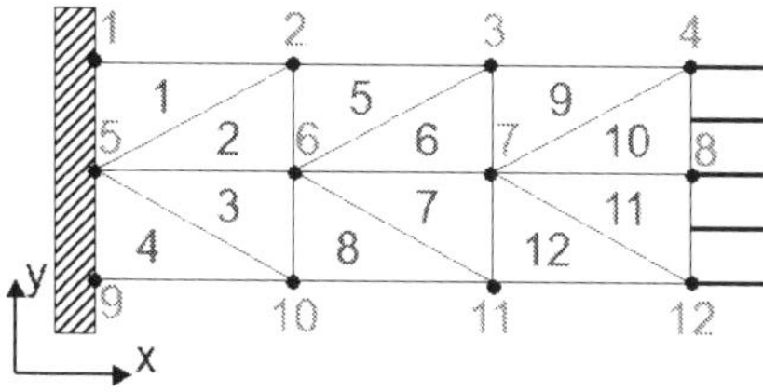

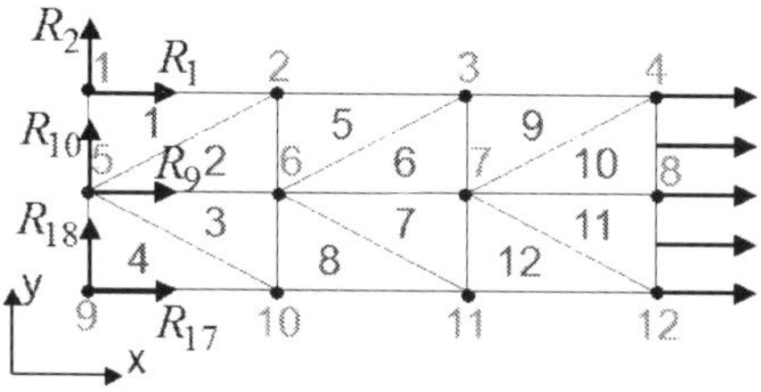

They don't disappear and should be included in the structural stiffness matrix.

We know that there will be unknown reactions at global nodes 1, 5 and 9. We could call these reactions R_1, R_2 , R_9, R_{10}, $R_{17}$and R_{18} (consistent with global displacements). So we have the free body of the structure:

Structural Equations of Equilibrium with Support Reactions

K	1	2	3	4	5	6	7	8	9	10	11	12	{q}		Q
1	X	X			X								$q_{1,2}$		$R_{1,2}$
2	X	X	X		X	X							$q_{3,4}$		
3		X	X	X		X	X						$q_{5,6}$		
4			X	X			X	X					$q_{7,8}$		X
5	X	X			X	X			X	X			$q_{9,10}$		$R_{9,10}$
6		X	X		X	X	X			X	X			=	
7			X	X		X	X	X			X	X			
8				X			X	X				X			X
9					X				X	X					$R_{17,18}$
10					X	X			X	X	X				
11						X	X			X	X	X			
12							X	X			X	X	$q_{23,24}$		X

X means that one or more elements have contributed here
** remember, each block is a (2x2) sub-matrix

OK, now one last step. We have to apply displacement boundary conditions. The structure is fixed at nodes 1, 5 and 9; thus, $q_1=q_2=q_9=q_{10}=q_{17}=q_{18}=0$. The easiest way to apply boundary conditions to any system of equations is as follows:

- Zero out the row and column on the left side matrix (the [K] matrix) corresponding to each B.C., and zero out the row of the right side (the {Q} matrix) corresponding to each B.C.
- Place a 1 on the diagonal of the left side matrix (the [K] matrix) corresponding to each B.C.
- You will notice that every dof that has a B.C. also corresponds to a dof where a support reaction (R) occurs. Applying B.C. as described above will thus eliminate the reactions from the equilibrium equations.
- A theoretical reason why we don't have to worry about reactions in structural equations of equilibrium? Because these support reactions R do no work (displacement is zero at support) and hence do not affect equilibrium of the structure!!!

Structural Equations of Equilibrium with B.C. Applied

K	1	2	3	4	5	6	7	8	9	10	11	12	{q}		Q
1	1	0			0								$q_{1,2}$		0
2	0	X	X		0	X							$q_{3,4}$		
3		X	X	X		X	X						$q_{5,6}$		
4			X	X			X	X					$q_{7,8}$		X
5	0	0			1	0			0	0			$q_{9,10}$		0
6		X	X		0	X	X			X	X			=	
7			X	X		X	X	X			X	X			
8				X			X	X				X			X
9					0				1	0					0
10					0	X			0	X	X				
11						X	X			X	X	X			
12							X	X			X	X	$q_{23,24}$		X

X means that one or more elements have contributed here
** remember, each block is a (2x2) sub-matrix

The structural equations with B.C. may now be solved for the unknown displacements. Note that when we solve the system of equations, the solution will give $q_1=q_2=q_9=q_{10}=q_{17}=q_{18}=0$, i.e, the 1st equation simply says (1) q_1=0, etc.

Element Strains and Stresses

Now we are ready to solve for the element strains and stresses. For each element, we can substitute the 6 global displacements

$$\underset{(3x1)}{\{\varepsilon^e\}} = \underset{(3x6)}{[B^e]}\underset{(6x1)}{\{q^e\}}$$

The stresses for each element can then be obtained

$$\underset{(3x1)}{\{\sigma^e\}} = \underset{(3x3)}{[D^e]}\underset{(3x1)}{\{\varepsilon^e\}}$$

Evaluation of stress results based on stress components in the Cartesian coordinates directions (σ_{xx}, σ_{yy}, σ_{xy}, etc) leaves something to be desired. Why? Stresses in these directions may not necessarily represent the largest stresses and we need these in order to consider yielding or failure. You already know that you can calculate principal stresses and maximum shear stress using stress transformation equations or Mohr's Circle. Hence, stress results (stress components) are often represented in two additional ways:

- Principal stresses and maximum shear stress, and
- von Mises stress.

Principal stresses can, as noted above, be obtained by either stress transformation equations or through the use of Mohr's Circle. An alternate approach to define principal stresses is to write:

$$\begin{vmatrix} \sigma_{xx}-\sigma_p & \sigma_{xy} & \sigma_{xy} \\ \sigma_{yx} & \sigma_{yy}-\sigma_p & \sigma_{yz} \\ \sigma_{zx} & \sigma_{zy} & \sigma_{zz}-\sigma_p \end{vmatrix} = 0$$

Expansion of the determinant provides a cubic equation that can be solved for the three principal stresses σ_p. Comparing principal stresses to a tensile yield stress provides some measure of evaluation; however, one has to keep in mind that comparing the principal stress (obtained from a three-dimensional stress state) to a yield stress obtained from a uniaxial tension test is risky at best.

The von Mises stress provides a means to extrapolate uniaxial tensile test data (for yield stress) to a three-dimensional stress state. In effect, the von Mises stress provides an "equivalent" uniaxial stress approximation to the three-dimensional stress state in a body through the following equation:

$$\sigma_{VM} < \sigma_y$$

Equation forms an ellipsoid in 3-D (ellipse in 2-D) when the stresses are plotted in principal stress space. As long as the stress state represented by the principal stresses is inside the ellipse (the yield surface), the material is elastic.

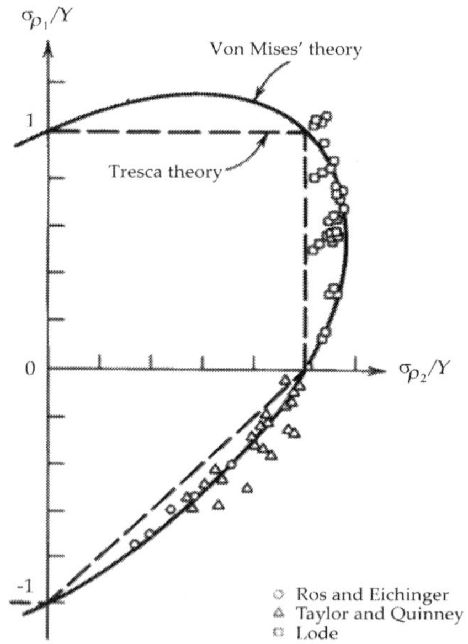

Figure 8.6: Experiment versus theory for yielding of copper, nickel, and steel in biaxial loading.

ELEMENT LIBRARIES

Choosing the right element for a structural component and loading, in order to maximize potential for correct results with the least amount of computation

Many, many finite elements have been developed for use in modern FEM software. Choosing the correct element for a particular structural is paramount. For example,

- if a structural member behaves like a beam in bending, we should choose a beam element to model it,
- if a structural member behaves like a thin plate in plane stress, we should choose an appropriate element to model it,
- if a structural member looks like a shell of revolution, we should use a thin shell of revolution element,
- if a structural member will experience a three-dimensional stress state, we have to choose an element that models that behavior, etc.
- Here are some examples of the types of elements available:
- Truss element (2-D and 3-D)
- Beam bending element (2-D and 3-D; straight and curved)

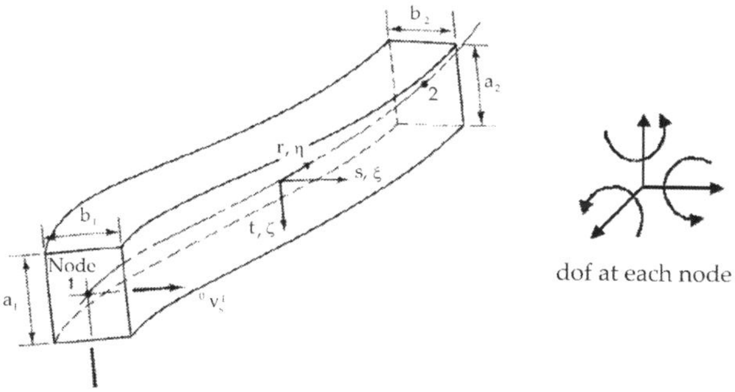

Figure 8.7: Beam bending elements.

- Membrane element (no bending; flat and curved)
- Triangular, Quad (both straight and curved sides)
- Planes Stress and Plane Strain elements
- Triangular and Quadrilateral shapes (both straight and curved sides).

Plane stress requires that the only non-zero stresses occur in the plane of the element (however, strain does occur normal to plane). Generally applicable to thin geometries. Two displacement dof per node (NO rotational dof).

Plane strain requires that the only non-zero strains occur in the plane of the element (strain is zero normal to plane, but stress is not zero). Long constrained geometries (for example, a long pipe, a dam). Elements with curved boundaries will always have 3 or more nodes per edge.

Figure 8.8: Plane elements.

- Plate and shell bending elements (bending and in-plane stresses; flat and curved elements)
- Triangular, Quad (both straight and curved sides)

Plate and shell bending elements are characterized as being thin compared to other dimensions, and having no stress normal to the plate (similar to plane stress).

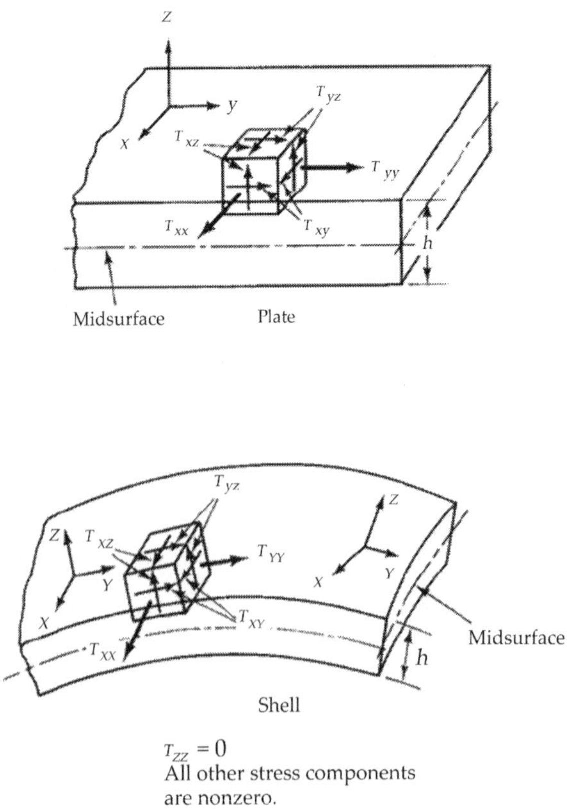

Figure 8.9: Comparison between plate and shell elements.

Plate and shell bending elements will have in-plane and normal displacements ( ) and rotations (θ_x, θ_y) about the two axes in the plane of the plate/shell. No stiffness about the normal axes.

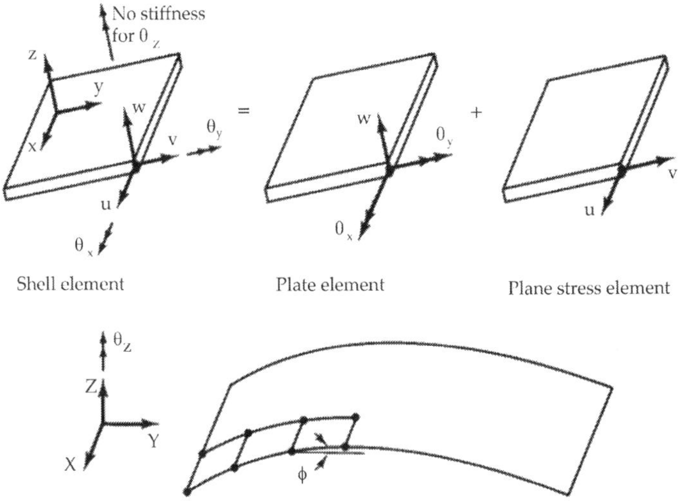

Figure 8.10: Use of a flat shell elements.

- Axisymmetric shell bending element (for shell of revolution)

The shell of revolution can be described geometrically by a curved (or straight) line that is revolved about the axis of symmetry. Top two figures are examples of thin shells of revolution. Degrees of freedom and stress state for a shell of revolution are similar to plate and shell bending elements, i.e., displacements parallel and perpendicular to the shell surface, and rotations about the two axes that lie in the plane of the shell. No stress normal to surface.

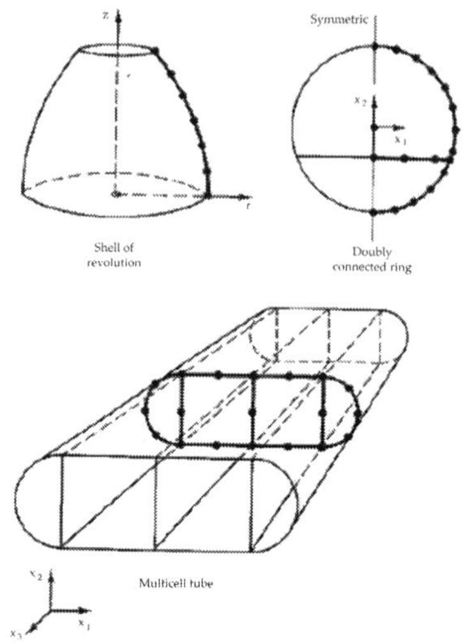

Figure 8.11: shell bending elements.

Lower figure (multicell tube) is NOT a shell of revolution, but would require plate or shell elements.

- Axisymmetric body of revolution element (3-D stress analysis)

Body is solid and axisymmetric about some axis of revolution. Stress state is fully three-dimensional (includes all stress components). Note that the element forms a triangular-shaped ring, i.e., the element itself is a body of revolution.

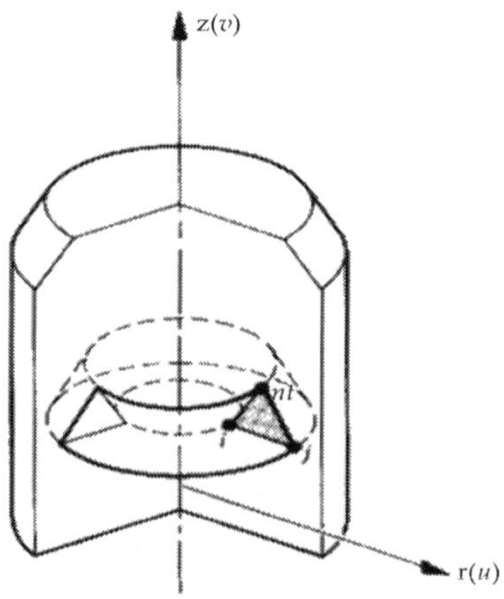

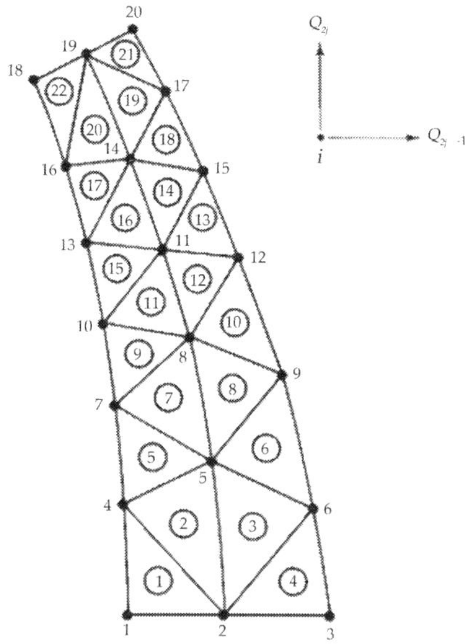

Figure 8.12: Element of an axisymmetric solid.

- General solid element (for full 3-D stress analysis)
- Tetrahedron, Brick shapes (both straight and curved sides)

Stress state is fully three-dimensional (includes all stress components). Three-dimensional solid elements have 3 displacements per node (NO rotational dofs). Elements with curved boundaries will always have 3 or more nodes per edge.

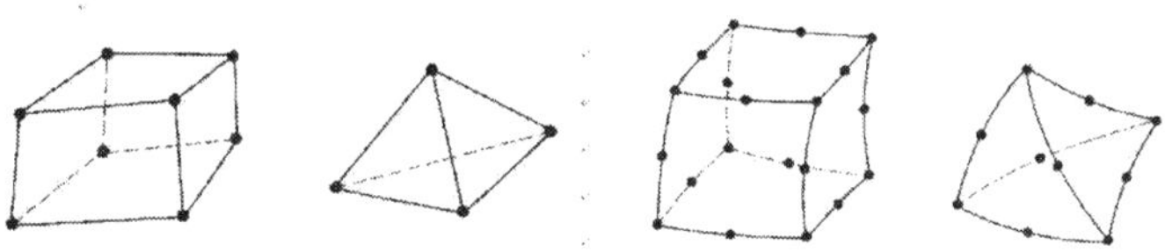

Figure 8.13: 3-D elements.

MESH GENERATION, PRE AND POST-PROCESSING, SOLVERS

In order to generate the structural stiffness matrix, we must specify the nodal coordinates (x, y, z) of each global node and element connectivity for each element. For a large problem with complex geometry, this is a monumental task if done by hand. Mesh generators are software tools that automate this process.

Mesh Generation and Preprocessing

In general, mesh generation starts by specifying the coordinates of a few key locations that will sufficiently define the outer boundary of a structural component and other distinguishing features such as holes, fillets, etc. Consider the bracket shown above. We could first define the corners of a 10" x 25" rectangle. To specify the rounded edges on the right boundary, we could specify that a 4" radius fillet is to be placed at each right corner. Finally, to specify the location of the hole, we would define a 2" radius circle whose center is located 5" from the right boundary.

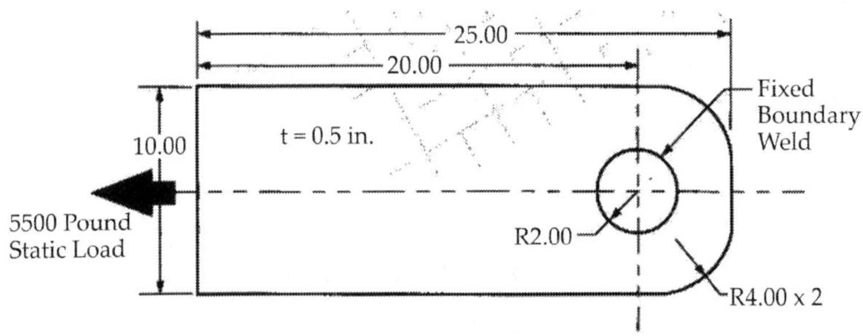

Figure 8.14: Problem description.

To specify the mesh (how many elements are to be used), we would specify so many elements in the x direction and so many elements in the y direction.

Since the structure is 2-D and in plane stress, we would specify the type of element (triangular, quad, etc.) to be used in defining the mesh.

The mesh generator would then automatically determine the coordinates of all global nodes and determine the element connectivity for all elements.

As a part of the mesh generation, we would also specify the thickness to be used for each element. Generally, the mesh generator would allow for a linear thickness variation within the region being meshed. For a variable thickness case, one would have to specify the thickness at several key locations and mesh generator would determine the thickness for all elements in the region.

Also, the material properties would be specified for each element in the region. Most mesh generators only provide for one material set to be used within a region (i.e., same material for all elements).

Next, the mesh generator (also called the *preprocessor*, because it processes all the required input) would require the type of loading that is applied to elements or element boundaries (point forces, body forces, surface tractions). For a thermal stress analysis, the preprocessor would also require information on thermal loading.

For a dynamic (transient) analysis, we would have to specify mass properties for each element and the time history of the loading.

Lastly, the displacement boundary conditions must be specified. Given all this information, the mesh generator/preprocessor will have generated a model something like that shown below:

With this information, the preprocessor would create a data set (an output file) in a format suitable to be read and processed by the finite element analysis program (also called the *solver*).

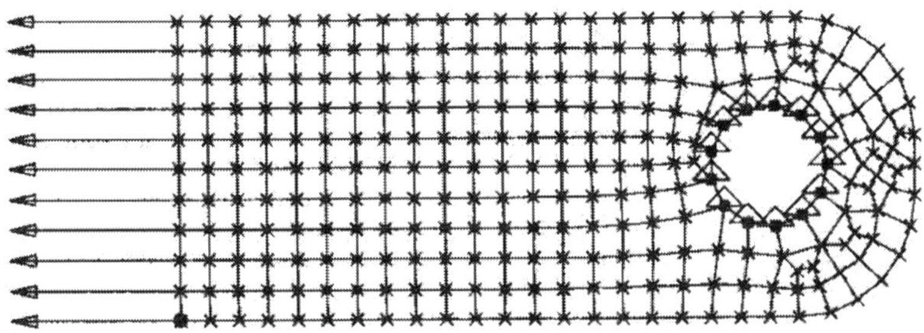

Figure 8.15: Define boundary condition.

SOLVER

Once the *solver* has created the structural stiffness and force matrices, solved for global nodal displacements, and solved for stresses and strains (plus a few other things appropriate to each element type), it generates an output file and we are now ready to examine and interpret the results.

Post-Processing

Clearly, for a complicated structure with many nodes and elements, the examination and evaluation of all results is an enormous task because of the shear volume of data (displacements, strains, stresses, etc.). The *post-processor* now takes over. Its purpose is to provide output such as the deformed geometry, contour plots of principal stress components, contour plots of von Mises stress, contour plots of strain components, etc. (what ever you decide is important). What, and how, the postprocessor can display information depends on the type of elements being used.

Often the pre and post-processor are combined into one software package (like FEMAP or PATRAN). Most preprocessor programs will create output files in formats acceptable to various solvers (like CAEFEM, NASTRAN, ANSYS, ABACUS, etc.). Likewise, most postprocessor programs will accept the output file of most solvers and display a variety of data results generated by the solver.

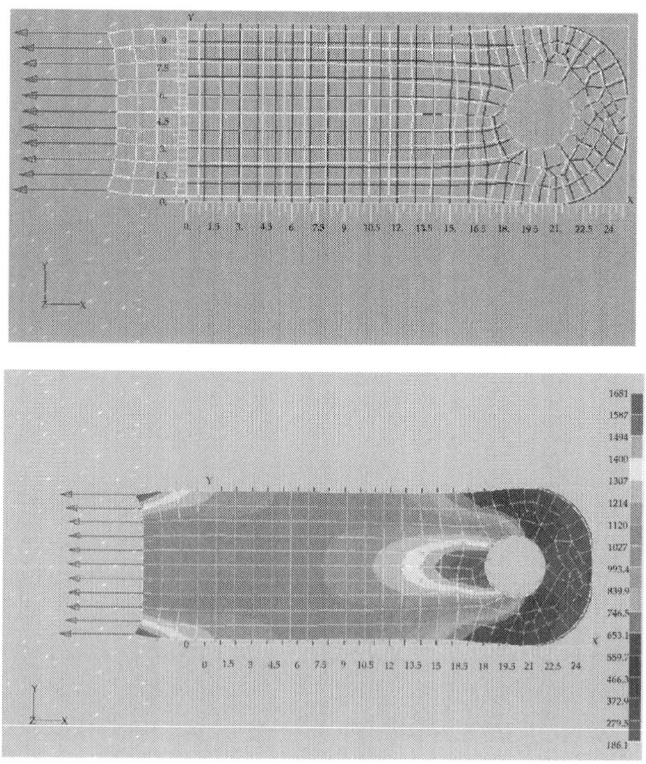

Figure 8.16: outputs of the deformed geometry and vonMises stress.

Last Thoughts on Mesh Generators, Pre/Post-Processors & Solvers

They will do a lot the tedious work for you.

They are not a substitute for good engineering know-how, judgement and experience.

In the end, YOU are responsible for the model you have generated for the structure, and for the correct interpretation and usage of the analysis results.

They aid you in the design process; but in the end, YOU are the designer.

REFERENCES

1. [AFS] Argyris, J.H., Fried, I. and Scharpf, D.W., The TUBA family of plate elements for the matrix displacement method. Aeronautical Journal of the Royal Aeron. Society, 72, 701-9, 1968.
2. [AH] Atamaz-Sibai, W. and Hinton, E., Adaptive mesh refinement with the Morley plate element. Proc. of NUMETA 90 Conference, Swansea, January 1990, 2, 1049-57, Elsevier Appl. Science, London, 1990.
3. [AHS] Atamaz-Sibai, W., Hinton, E. and Selman, A., Adaptive mesh refinement with the Mindlin-Reissner elements. Proc. of 2nd Int. Conf. an Computer Aided Analysis and Design of Concrete Structures., Zell-an-see, Austria, Abril 1990, Pineridge Press, 1990.
4. [AK] Argyris, J.H. and Kelsey, S. Energy theorems and structural analysis. Butterworth, London, 1960 (originally published in a series of articles in Aircraft Engineering, 1954-55).
5. [ASME] ASME. Guide for Verification and Validation in

Computational Solid Mechanics. ASME V& V 10-2006 American Society of Mechanical Engineers, 2006. Available at http://catalog.asme.org/Codes/PrintBook/VV 10 2006 Guide Verification.cfm

6. [Ba] Baker, T.J., Automatic mesh generation for complex three-dimensional regions using a constrained Delaunay triangulation. Engineering Computations, 5, 161-175, 1989.
7. [Bat] Bathe, K.J., Finite element procedures. Prentice Hall, Inc., 1996.
8. [BB] Belytschko, T. and Bachrach, W.E., The e±cient implementation of quadrilaterals with high coarse mesh accuracy. Comput. Meth. in Appl. Mech. Engrg, 54, 276-301, 1986.
9. [BCHC+] Barbat A.H., Cervera M., Hanganu A., Cirauqui C. and O~nate E., Failure pressure evaluation of the containment building of a large dry nuclear power plant. Nuclear Engineering and Design, 180, 251-270, 1998.
10. [BCF] Boussetta, R., Coupez, T. and Fourment, L., Adaptive remeshing based on a posteriori error estimation for forging simulation. Comput. Meth. in Appl. Mech. Engrg, 155, 6626-6645, 2006.
11. [BD] Batoz, J.L. and Dhatt, G., Modelisation des structures par elements finis. Vol. 1: Solides elastiques. Hermes, Paris, 1990.
12. [Be] Bedrosian, G., Shape functions and integration formulas for three-dimensional finite element analysis. Int. J. Num. Meth. Engrg., 35(1), 95-108, 1992.
13. [Bey] Beyer, W.H., CRC Standard Mathematical Tables. CRC Press, 28th ed., 1988.
14. [Bl] Blacker, T.D., FASTQ Users Manual Version 1.2. Technical Report SAND88-1326, Sandia National Laboratories, Albuquerque, New Mexico, 1994.
15. [Bl2] Blacker, T.D., Automated quadrilateral surface discretization method and apparatus usable to generate mesh in a finite element analysis system. Unites States Patent, 5,315,537, May 1994.
16. [BM] Blacker, T.D. and Meyers, R.J., Seams and wedges in plastering: A 3D hexahedral mesh generation algorithm. Engineering with Computers, 2(9), pp. 83-93, 1993.

17. [Bo] Bowyer, A., Computing Dirichlet tessellations. The Computer Journal, 24(2), 162-166, 1981.
18. [BO] Bugeda, G. and Oliver, J., A general methodology for structural shape optimization problems using automatic adaptive remeshing. Int. J. Num. Meth, Engrg., 36(18), 3161-3185, 1993.
19. [BO2] Bugeda, G. and Onate, E., New adaptative technique for structural problems. First European Conference on Numerical Methods in Engineering, Ch. Hirsch, O.C. Zienkiewicz and E. Onate (Eds.), pp. 135{142, Elsevier, 7-11 September 1992.
20. [BO3] Bugeda, G. and Onate, E., Adaptive techniques for elliptic problems. In Advances in Post and Preprocessing for Finite Element Technology, Papadrakakis M. and Topping B.H.V. (Eds.), Civil-Comp Press, 77-84, 1994.
21. [BP] Bonet, J. and Peraire, J., An alternate digital tree algorithm for geometric searching and intersection problems. Int. J. Num. Meth, Engrg., 31, 1-17, 1991.
22. [BS] Blacker, T.D. and Stephenson, M.B., Paving: A new approach to automated quadrilateral mesh generation. Int. J. Num. Meth, Engrg., 32, 811-847, 1991.
23. [BS2] Brenner S.C. and Scott L.R., The mathematical theory of finite element methods. SpringerVerlag, 1994.
24. [BSC] Blacker, T.D., Stephenson, M.B. and Canann, S.A., Analysis automation with paving: A new quadrilateral meshing technique. Advances in Engineering Software, 56, 332-337, 1991.
25. [BSGU] Babu¸ska, I., Stroubolis, T., Gangaraj, S.K. and Upadhyay, C.S., Pollu-tion error in the h-version of the finite element method and the local quality of the recovered derivatives. Comput. Meth. in Appl. Mech. Engrg, 140, pp. 1-37, 1997.
26. [BSUG] Babu¸ska, I., Stroubolis, T., Upadhyay, C.S. and Gangaraj, S.K., A posteriori estimation and adaptive control of the pollution error in the h-version of the finite element method. Int. J. Num. Meth, Engrg., 38, pp. 4207-4235, 1995.
27. [Bu] Bugeda, G., Error estimation technique and mesh generation procedure in structural shape optimization, Ph.D. Thesis (in spanish), Civil Engineering School, Technical University of Catalonia (UPC),

Barcelona, 1990.

28. [Bu2] Bugeda, G., A comparison between new adaptive remeshing strategies based on point wise stress error estimation and energy norm error estimation, Communications in Numerical Methods in Engineering, 18, pp. 469{-482, 2002.
29. [BWSG+] Baehmann, P.L., Wittchen, S.L., Shephard, M.S., Grice, K.R. and Yerry, M.A., Robust geometrically-based, automatic two-dimensional mesh generation. Int. J. Num. Meth. Engrg., 24, 1043-1078, 1987.
30. [BY] Boissonnat, J.D. and Yvinec, M., Algorithmic Geometry. Cambridge Uni-versity Press, 1997.
31. [BZ] Boroomand, B. and Zienkiewicz, O.C., Recovery by equilibrium patches (REP). Int. J. Num. Meth, Engrg., 40, 137-154, 1997.
32. [BZ2] Boroomand, B. and Zienkiewicz, O.C., An improved REP recovery and the effectivity robustness test. Int. J. Num. Meth, Engrg., 40, 3247-3277, 1997.
33. [Ca] Carey, G.F., Computational Grids: Generation, Adaptation and Solution Strategies. Taylor and Francis, 1997.
34. [CB] Coorevits, P. and Bellenger, E., Alternative mesh optimality criteria for h¡adaptive ¯nite element method. Finite Element in Analysis and Design,40, 1195-1215, 2004.
35. [CBMB] Cass, R.J., Benzley, S.E., Meyers, R.J. and Blacker, T.D., Generalized 3D paving: An automated quadrilateral surface mesh generation algorithm.Int. J. Num. Meth. Engrg., 39, 1475-1489, 1996.
36. [CDOO] Celentano, D., Dabir, A., Oller, S. and Onate, E., Thermal-mechanical analysis of a S.G. cast iron crankshaft during cooling and solidification in green molding sand. Technical Report CIMNE, IT 182, November 1995.
37. [CFF] Cavendish, J.C., Field, D.A. and Frey, W.H., An approach to automatic three dimensional finite element mesh generation. Int. J. Num. Meth, Engrg., 21, 329-347, 1985.
38. [CI] Calvo, N.A. and Idelsohn, S.R., All-hexahedral element meshing: Gener-ation of the dual mesh by recurrent subdivision.

Comput. Meth. in Appl. Mech. Engrg, 182, pp. 371-378, 2000.

39. [Cl] Clough, R.W., The finite element method in plane stress analysis. Proc. 2nd A.S.C.E. Conf. in Electronic Computation, Pittsburgh, Pa., Sept. 1960.
40. [Cl2] Clough, R.W., Comparison of three dimensional finite elements. Proceed-ings Symposium on Applications of Finite Element Methods in Civil En-gineering, Vanderbilt Univ., Nashville, Tenn., (published by ASCE), 1-26, 1969.[CLT] Cantin, G., Loubignac, G. and Touzot, G., An iterative algorithm to build continous stress and displacement solutions. *Int. J. Num. Meth. Engrg.*, 12, 1493-1506, 1978.
41. [CMOS] Codina, R., Morton, C. Onate, E. and Soto, O., Numerical aerodynamic analysis of large buildings using a finite element model with application to a telescope building. Int. Journal of Numerical Methods for Heat and Fluid Flow, 10(6), 616-633, 2000.
42. [CMPW] Cook, R.D., Malkus, D.S., Plesha, M.E. and Witt, R.J., Concepts and application of finite element analysis. 4th ed., John Wiley and Sons, 2002.
43. [Co] Courant, R., Variational methods for the solution of problems of equilibrium and vibration. Bull. Am. Math. Soc., 49, 1-23, 1943.
44. [Co2] Courtney, T.H., Mechanical behaviour of materials. McGraw-Hill, 1990.
45. [COG] Cervera, M., Oliver, J. and Galindo, M., Numerical simulation of damage in concrete dams (in Spanish), Monograph N. 4, CIMNE, Barcelona, 1991.
46. [COHO] Cervera, M., Oliver, J., Herrero, E. and Onate E., A computational model for progressive cracking in large dams due to swelling of concrete. Int. Journal of Fracture, 35(1,2,3), 573-85, 1990.
47. [COHR+] Casteleiro, M., O~nate, E., Huerta, A., Roig, J. and Alonso, E., Three dimensional analysis of no tension materials. 4th. Int. Conf. on Rock Mech, Melbourne, Australia, 1983.
48. [Cow] Cowper, G.R., Gaussian quadrature formulae for triangles. *Int. J. Num. Meth. Engrg.*, 7, 405-8, 1973.
49. [Cr] Crisfield, M.A., Finite element and solution procedures for

structural anal-ysis, I: Linear analysis. Pineridge Press, 1986.

50. [CVDO+] Celentano, D., Visconte, D., Dardati, P., Oller, S. and Onate, E., A thermomechanical model for solidification problems. Experimental validation. In Proceedings of Fourth International Conference on Computational Plasticity (COMPLAS IV), R. Owen, E. Onate and E. Hinton (Eds.), Barcelona, Spain, 3{6 April 1995, Pineridge Press/CIMNE, pp. 2385-2396.
51. [Dao] de Arantes Oliveira, E.R., The patch test and the general convergence criteria of the finite element method. Int. J. of Solids and Structures, 13, 159-178, 1977.
52. [De] Delaunay, B., Sur la sphμere vide. Izv. Akad. Nauk SSSR, Otdelenie Matematicheskii i Estestvennyka Nauk, 7, 793-800, 1934.
53. [Dem] Demmel, J., Applied Numerical Linear Algebra. Society for Industrial and Applied Mathematics, Philadelphia, PA, 1997.
54. [DH] Davis, P.J. and Hersch, R., The Dream of Descartes. Ed. Labor, 1989.
55. [DH2] Diez, P. and Huerta, A., A unified approach to remeshing strategies for finite element h-adaptivity. Comput. Meth. in Appl. Mech. Engrg, 176(1-4), 215-229, 1999.
56. [Dho] Dhondt, G., A new automatic hexahedral mesher based on cutting. Int. J. Num. Meth, Engrg., 50, 2109-2126, 2001.
57. [Du] Dumavanet, D.A., High degree efficient symmetrical Gaussian quadrature rules for the triangle. Int. J. Num. Meth. Engrg., 21, 7, 1985.
58. [DV] Dvorkin, E.N. and Vassolo, S.I. A quadrilateral 2D finite element based on mixed interpolation of tensorial components. Engineering Computations, 6, 217-224, 1989.
59. [EIZ] Ergatoudis, J.G., Irons, B.M. and Zienkiewicz, O.C., Curved, isoparamet-ric, quadrilateral elements for finite element analysis. Int. J. of Solids and Structures, 4, 31-42, 1968.
60. [Fa] Farin, G., Curves and Surfaces for Computer Aided Geometric Design. Academic Press, 1990.
61. [Fa2] Farin, G., Curves and Surfaces for CAGD. A Practical Guide. 5th Edition, Morgan Kaufmann Publishers Inc., San Francisco, CA, USA, pp. 499, 2001.

62. [FdV] Fraeijs de Veubeke, B., Variational principles and the patch test. Int. J. Num. Meth. Engrg., 8, 783-801, 1974.

63. [Fe] Ferrers, N.M., An elementary treatise on trilinear coordinates, the method of reciprocal polars and the theory of projections. MacMiilan , London, 1961.

64. [Fe2] Ferencz, R.M., Element-by-element preconditioning techniques for large-scale, vectorized finite element analysis in nonlinear solid and structural me-chanics. Ph.D thesis, Department of Mechanical Engineering, Stanford University, Stanford, California, 1989.

65. [FG] Frey, P.J. and George, P.L., Mesh Generation Application to Finite Elements. Hermes Science Publishing, Oxford, Paris, 816 , 2000.

66. [FNS] FrÄoier, M., Nilsson, L. and Samuelsson, A., The rectangular plane stress element by Turner, Pian and Wilson. Int. J. Num. Meth. Engrg., 8, 2, 433-7, 1974.

67. [FP] Faux, I.D. and Pratt, M.J., Computational Geometry for Design and Manufauture. Ellis Horwood, 1985.

68. [FO] Flores, R. and Ortega, E., Pumi: An explicit 3D unstructured finite element solver for the Euler equations. Publication CIMNE PI 326, November 2008.

69. [FO2] Flores, R., Ortega, E. and Onate, E., A numerical study of wing deformation effects in wind tunnel testing using twin-sting supports. Publication CIMNE, 2008. Submitted to Engineering Computations.

70. [Fu] Fourment, L., Notes on the mesh optimality criteria of Onate and Bugeda [OB]. Private Communication, 1995.

71. [Ga] Gallagher, R.H., Finite Element Analysis Fundamentals. Prentice{Hall, Englewood Cliffs, N.J., 1975.

72. [GB] George, P.L. and Borouchaki, H., Delaunay Triangulation and Meshing. Application to Finite Elements. Editions Hermes, Paris, 1998.

73. [GBFL+] George, P.L., Borouchaki, H., Frey, P.J., Laug, P. and Saltel, E., Mesh Generation and Mesh Adaptivity. In Encyclopedia of Computational Mechanics. Chapter 17, 1 Fundamentals, E. Stein,

R. De Borst, J.R. Thomas (Eds.), J. Wiley, 2004.

74. [Ge] George, P.L., Automatic mesh generation. Applications to the finite element method. J. Wiley, 1991.
75. [GH] Geller, M. and Harbord, R., Moderate degree cubature formulas for 3D tetrahedral finite element approximations. Communications in Applied Nu-merical Methods, 7(6), 487-495, 1991.
76. [GH2] Gordon, W.J. and Hall, C.A., Construction of curvilinear co-ordinate systems and application to mesh generation. Int. J. Num. Meth, Engrg., 3, 461-477, 1973.
77. [GHS] George, P.L., Hecht, F. and Saltel, E., Automatic mesh generator with specified boundary. Comput. Meth. in Appl. Mech. Engrg, 92, 269{288, 1991.
78. [GiD] GiD. The personal pre and postprocessor. CIMNE, Barcelona, 2008, www.gidhome.com.
79. [GS] George, P.L. and Seveno, E., The advancing-front mesh generation method revisited. Int. J. Num. Meth, Engrg., 37, 3605-3619, 1994.
80. [HC] Hinton, E. and Cambell, J.S., Local and global smoothing of discontinous element functions using a least square method. Int. J. Num. Meth. Engrg., 8, 461-80, 1979.
81. [HCA] Huerta, A., Casteleiro, M. and Alonso, E., Non linear numerical analysis of earth dam constructions, in Numerical Methods for non Linear Problems, C. Taylor, E. Hinton, R. Owen y E. Onte (Eds.), Pineridge Press, 1984.
82. [HCB] Hughes, T.J.R., Cottrell, J.A. and Bazilevs, Y., Isogeometric analysis: CAD, finite elements, NURBS, exact geometry and mesh refinement. Com-put. Meth. in Appl. Mech. Engrg, 194, 4135-4195, 2005.
83. [He] Hearmon, R.F.S., An introduction to applied anisotropic elasticity. Oxford Univ. Press, 1961.
84. [HH] Hooley, R.F. and Hibbert, P.D., Bounding plane stress solutions by finite elements. Journal Struct. Division, ASCE, 92, ST1, 39-48, 1966.
85. [HKD] Hitchings, D., Kamoulakos, A. and Davies, G.A.O., Linear Static Bench-marks. Vol 1, National Engineering Laboratory Report,

January 1987 NAFEMS LSB1.

86. [HO] Hinton, E. and Owen, D.R.J., Finite element programming. Academic Press, 1979.

87. [HO2] Hinton, E. and Owen, D.R.J., Introduction to finite element computations. Pineridge Press, 1980.

88. [Hr] Hreniko, A., Solution of problems in elasticity by the framework method. J. Appl. Mech., 18, 169{75, 1991.

89. [Hu] Hughes, T.J.R., The finite element method. Linear static and dynamic finite element analysis. Prentice-Hall, 1987.

90. [IA] Irons, B.M. and Ahmad, S., Techniques of finite elements. Ellis Harwood, Chichester, 1980.

91. [Ir] Irons, B.M., Numerical integration applied to finite element method. Conf. Use of Digital Computers in Struct. Eng., Univ. Newcastle, 1966.

92. [IR] Irons, B.M. and Razzaque, A., Experience with the patch test for convergence of finite element method. The Mathematics of Finite Elements with Application to Partial Differential Equations, 557{587, A.R. Aziz (ed.), Academic Press, 1972.

93. [Joe] Joe, B., Quadrilateral mesh generation in polygonal regions. Computer Aided Design, 27, 209{222, 1995.

94. [Jor] Jordan, W.B., The plane isoparametric structural element. General Electric Co, Report KAPL{M{7112 , Schenectadly, New York, 1970.

95. [JSK] Johnston, B.P., Sullivan, J.M. and Kwasnik, A., Automatic conversion of triangular element meshes to quadrilateral elements. Int. J. Num. Meth. Engrg., 31, 67{84, 1991.

96. [JT] Jin, H. and Tanner, R.I., Generation of unstructured tetrahedral meshes by advancing front technique. Int. J. Num. Meth. Engrg., 36, 1805{1823, 1993.

97. [JW] Jin, H. and Wiberg, N.E., Two dimensional mesh generation, adaptive remeshing and refinement. Int. J. Num. Meth. Engrg., 29(7), 1501{1526, 1990.

98. [KF] Koslo, D. and Frasier, G.A., Treatment of hour glass patterns in low order finite element codes. Int. J. for Numerical Analysis Methods in Geomechanics, 2, 57{72, 1978.

99. [KGZB] Kelly, D.W., De S.R. Gago, J.P., Zienkiewicz, O.C. and Babuska, I., A posteriori error analysis and adaptive processes in the ¯nite element method: Part I: Error analysis. Int. J. Num. Meth, Engrg., 19, 1593-1619, 1983.

100. [KHK] Kwok, W., Haghighi, K. and Kang, E., An effcient data structure for the advancing-front triangular mesh generation technique. Communications in Numerical Methods in Engineering, 11(5), 465-473, 1995.

101. [KS] Knupp, P. and Steinberg, S., The Fundamentals of Grid Generation. CRC Press, 1993.

102. [LB] Lee, N.-S. and Bathe, K.-J., Effects of element distortion on the perfor-mance of isoparametric elements. Int. J. Num. Meth. Engrg., 36, 3553-3576, 1993.

103. [LB2] Li, L.Y. and Bettess, P., Notes on mesh optimal criteria in adaptive finite element computation. Communications in Numerical Methods Engineering,

104. 11(11), 911-915, 1995.

105. [LB3] Li, L.Y. and Bettess, P., Adaptive finite element methods: a review. Appl. Mech. Reviews, 50(10), 581-591, 1997.

106. [LBBB] Li, L.Y., Bettess, P., Bull, J., Bond, T. and Ipplegarth, I., Theoretical formulations for adaptive finite element computations. Communications in Numerical Methods Engineering, 11(10), 857-868, 1995.

107. [Le] Lekhnitskii, S.G., Theory of elasticity of an anisotropic elastic body. Holden Day, San Francisco, 1963.

108. [Li] Livesley, R.K., Matrix methods in structural analysis. 2nd ed., Pergamon Press, 1975.

109. [LJ] Liu, A. and Joe, B., Relationship between tetrahedron shape measures. BIT, 34, 268-287, 1994.

110. [Lo] Lo, S.H., A new mesh generation scheme for arbitrary planar domains. Int. J. Num. Meth, Engrg., 21, 1403-1426, 1985.

111. [LP] LÄohner, R. and Parikh, P., Three-dimensional grid generation by the ad-vancing front method. Int. J. Num. Meth. Engrg., 8, 1135-1149, 1988.

112. [LP2] Ladeveze, P. and Pelle, J.P., Mastering calculations in linear

and non linear mechanics. Springer, 2005.

113. [LTV] Lober, R.R., Tautges, T.J. and Vaugham, C.T., Parallel paving: An algo-rithm for generating distributed, adaptive, all-quadrilateral meshes on paral-lel computers. Technical Report SAND97-0545, Sandia National Laboratory, Alburquerque, New Mexico, USA, 1997.

114. [LW] Li, X.D. and Wiberg, N.-E., A posteriori error estimate by element patch postprocessing, adaptive analysis in energy and L2 norm. Computers and Structures, 53, 907-919, 1994.

115. [LZG] Lewis, R.W., Zheng, Y. and Gethin, D.T., Three-dimensional unstruc-tured mesh generation: Part 3. Volume meshes. Comput. Meth. in Appl. Mech. Engrg, 134, 285-310, 1996.

116. [MBBO] Miquel, J., Botello, S., Buil, J. and Onate, E., The Talvachia double curvature arch dam. Static and dynamic analysis (in spanish). Monograph No. 6, CIMNE, Barcelona, 1991.

117. [Mc] Mc. Henry, D., A lattice analogy for the solution of plane stress problems.J. Inst. Civ. Engng., 21, 59-82, 1943.

118. [MDR] Mangalgiri, P.D., Dattaguru, B. and Rammamurthy, T.S., Specification of skew conditions in finite element formulation. Int. J. Num. Meth. Engrg., 12(6), 1037-1041, 1978.

119. [MH] Moller, P. and Hansbo P., On advancing front mesh generation in three dimensions. Int. J. Num. Meth. Engrg., 38, 3551-3569 1995.

120. [MW] Marcum, D.L. and Weatherill, N.P., Unstructured grid generation using iterative point insertion and local reconnection. AIAA Journal, 33(9), pp. 1619-1625, 1995.

121. [Na] Narayanaswani, R., Dependence of plate bending finite element defections and eigenvalues on Poisson's ratio. A.I.A.A.J., 12. 1420-21, 1974.

122. [Ng] Nguyen, V.P., Automatic mesh generation with tetraedon elements. Int. J. Num. Meth. Engrg., 18, 273{89, 1982.

123. [No] Noor, A., Bibliography of books and monographs on finite element tech-nology. Appl. Mech. Rev., 44(6), 307-17, June, 1991.

124. [OB] Onate, E. and Bugeda, G., A study of mesh optimality criteria in adaptive finite element analysis. Engineering Computations, 10(4), 307-321 1993.

125. [OC] Onate, E. and Castro, J., Adaptive mesh refinement techniques for struc-tural problems. In The Finite Element Method in the 1990's, A book dedi-cated to O.C. Zienkiewicz , E. Onate, J. Periaux and A. Samuelsson (Eds.), Springer-Verlag/CIMNE, Barcelona, 1991.

126. [OCCO+] Onate, E., Celentano, D., Codina, R., Oller, S. and Soto, O., Finite element model for analysis of mould filling and solidification in casting. In Proceedings of the Conference on Modelling of Casting, Welding and Advanced Solidification Processes VII, London, 10-15 September, 1995.

127. [OCK] Onate, E., Castro, J. and Kreiner, R., Error estimation and mesh adaptivity techniques for plate and shell problems. In The 3rd. International Conference on Quality Assurance and Standars in Finite Element Methods, NAFEMS, Stratford-upon-Avon, England, 10-12 September, 1991.

128. [OCOH] Oliver, J., Cervera, M., Onate, E. and Herrero, E., A case study of gravity dam subjected to severe internal actions: Description and numerical simulations. In Sixth Congress on Large Dams. International Committee for Large Dams (ICOLD), San Francisco, C.29, 1387-409, 1988.

129. [OGI] Onate, E., Garcia, J. and Idelsohn, S.R., Ship hydrodynamics. Encyclo-pedia of Comput. Mechanics, John Wiley & Sons Ltd, Vol. 3, Chapter 18, 579-607, 2004.

130. [OH] Onate, E. and Hanganu, A., Advanced method for the analysis of the failure load of concrete structures (in Spanish). Publication CIMNE, PI 176, September 1999.

131. [OHBO+] Onate, E., Hanganu, A., Barbat, A., Oller, S., Vitaliani, R., Saetta, A. and Scotta, R., Structural analysis and durability assessment of histori-cal constructions using a finite element damage model. Publication CIMNE, P173, pp. 26, 1995.

132. [OHBO+2] Onate, E., Hanganu, A., Barbat, A., Oller, S., Vitaliani, R., Saetta, A. and Scotta, R., Structural Analysis and durability assessment of histori-cal constructions using a finite element damage model. Publication CIMNE, P173, 1995. Also in Structural Analysis of Historical Constructions. Possibil-ities of Numerical and Experimental Techniques, Roca et al. (Eds.), CIMNE, Barcelona, 1997.

133. [On] Onate, E., Reliability analysis of concrete structures. Numerical and ex-perimental studies. Publication CIMNE, PI 107, January 1997.

134. [On2] Onate, E., Structural Analysis with the Finite Element Method. Linear Statics. Volume 2: Beams, Plates and Shells. CIMNE-Springer, 2009.

135. [OOB] Onate, E., Oliver, J. and Bugeda, G., Finite element analysis of the non linear response of concrete dams subjected to internal loading. Finite Element Methods for Non Linear Problems, Europe{US Symposium, Trodheim, Norway, 1985, P. Bergan et al. (Eds.), Springer-Verlag, 1985.

136. [OPSE] Onate, E., Pasenau, M., Suit, J. and Escolano, E., New paradigms and tools for interfacing geometrical models, analysis data, numerical methods and visualization software in computational science and engineering: The GiD project. Publication CIMNE, PI 262, May 2005.

137. [Or] O Rourke, J., Computational Geometry in C. Cambridge University Press, 2nd Edition, 2001.

138. [OSCS] Owen, S.J., Staten, M.L., Canann, S.A. and Saigal, S., Q-morph: An in-direct approach to advancing front quad meshing. Int. J. Num. Meth. Engrg., 44, 1317-1340, 1999.

139. [Pe] Peiro, J., A finite element procedure for the solution of the Euler equa-tions on unstructured meshes. Ph.D. Thesis, Civil Eng. Dpt., Univ. College of Swansea, U.K., 1989.

140. [Pe2] Peiro, J., Surface grid generation. In Handbook of Grid Generation, Chap-ter 19, 19.1-19.20, CRC Press, 1999.

141. [PFGS] Pereda, J., Foces, A., Garrido, J.A. and Suarez, S., Multicellular box docks: An user friendly Internet based software for design in civil engineering. 2nd Conference on Advances and Applications of GiD, Publication CIMNE No. 210, Barcelona, 2004.

142. [PFTV] Pres, W.H., Flannery, B.P., Teukolsky, S.A. and Vetterling, W.T., Nu-merical Recipes. The art of Scienti¯c Computing. Cambridge Univ. Press, 1986.

143. [PGH] Parthasarathy, V.N., Graichen, C.M. and Hathaway, A.F., A comparison of tetrahedron quality measures. Finite Elements in Analysis and Design, 15, 255-261, 1993.

144. [Pi] Pissanetzky, S., Kubik: An automatic three dimensional mesh generator.
145. Int. J. Num. Meth. Engrg., 17(2), 255-69, 1981.
146. [PP] Prinja, N.K. and Puri, A.K., An Introduction to the Use of Material Models in FE. NAFEMS, November 2005.
147. [PPFM+] Peraire, J., Peiro, J., Formaggio, L., Morgan, K.J. and Zienkiewicz, O.C., Finite element Euler computations in three dimensions. Int. J. Num. Meth. Engrg., 26, 2135-2159, 1988.
148. [PPM] Peraire, J., Peiro, J. and Morgan, K., Adaptive remeshing for 3-dimensional compressible °ow computations. Journal of Computational Physics, 103, 269-285, 1992.
149. [Pr] Przemienieck, J.S., Theory of matrix structural analysis. McGraw-Hill, New York, 1968.
150. [PS] Preparata, F.P. and Shamos, M.I., Computational Geometry. Springer-Verlag, New York, 1988.
151. [PT] Piegl, L. and Tiller, W., The NURBS Book. 2nd Edition, Springer-Verlag, New York, 646, 1997.
152. [PVMZ] Peraire, J., Vahdati, M., Morgan, K. and Zienkiewicz, O.C., Adaptive remeshing for compressible flow computation. J. Comp. Phys., 72, 449-66, 1987.
153. [Rad] Radau, Journal of Math., 3, p. 283, 1880.
154. [Ral] Ralston, A., A First Course in Numerical Analysis. McGrawHill, New York, 1965.
155. [RCMR+] Roca, P., Casarin, F., Modena, C., Rodr¶³guez, I. and Garay, A., Damage monitoring of long-span historical structures. In Damage of massive historical constructions, L. Vinda (ed.), WIT Press, Southampton, UK, 2006.
156. [Re] Rebay, S., Efficient unstructured mesh generation by means of Delau-nay triangulation and Bowyer-Watson algorithm. Journal of Computational Physics, 106, 125-138, 1993.
157. [REMFI] REMFI: Rear fuselage and empennage flow investigation. Project No. AST3-CT-2004-502895, sponsored by the FP6 Programme of the European Commission (EC), 3/1/2004{2/28/2007, www.cimne.com/remfi/.
158. [RG] Rosanoff, R.A. and Ginisburg, T.A., Matrix error analysis for

engineers.

159. Proc. First Conf. on Matrix Meth in Struct. Mech., Wright-Patterson, Air Force B., Ohio, 1965.

160. [RGAA] Roca, P., Gonzalez, J.L., Aguerri, F. and Aguerri, J.I., Monitoring of long-term damage in Gothic Cathedrals. In Structural Studies, Repairs and Maintenance of Heritage Architecture VII (STREMAH 7), C. Brebbia (Ed.), WIT Press, Southampton, pp. 109-115, Halkidiki, Grecia, 2003.

161. [RGL] Rosanoff, R.A., Gloudemann, J.F. and Levy, S., Numerical condition-ing of stiffness matrix formulations for frame structures. Proc. Second Conf. on Matrix Meth. in Struct. Mech., Wright-Pattersson, AFB, Ohio, 1968 (AFFDL-TR-68-150. Dec, 1969: AD-703-685 N.T.T.S) 1029-60.

162. [RGMO] Roca, P., Gonzalez, J.L., Mari, A.R. and Onate, E. (Eds.), Structural Analysis of Historical Constructions I. Possibilities of Numerical and Exper-imental Techniques. CIMNE, Barcelona, 1997.

163. [RGOL] Roca, P., Gonzalez, J.L., O~nate, E. and Lourenco, P.B. (Eds.), Struc-tural Analysis of Historical Constructions II. Possibilities of Numerical and Experimental Techniques. CIMNE, Barcelona, 1998.

164. [RM] Roca, P. and Molins, C., Structural assessment of the cimborios of two Gothic Cathedrals. In IASS-MSU International Symposium. Bridging Large Spans: From Antiquity to the Present, Sanayi-i-Nefise, pp. 145-154, Istambul, Turquia, 2000.

165. [Ro] Roache, P.J., Verification and Validation in Computational Science and Engineering. Hermosa, Albuquerque, 1998.

166. [Ro2] Roca, P., Structural Monitoring of three Gothic Cathedrals. In On-site control and evaluation of masonry structures, RILEM Publication, L. Vinda (Ed.), pp. 105-118, Bagneux, Francia, 2003.

167. [ROD] Rachowicz, W., Oden, J.T. and Demkowicz, L., Toward a universal h ¡p adaptive finite element strategy. Part 3: Design of h-p meshes. Comput. Meth. in Appl. Mech. Engrg, 77, 181-211, 1989.

168. [RPOH] Roca, P., Pellegrini, L., Onate, E. and Hanganu, A., Analysis of the structure of Gothic Cathedrals. Application to Barcelona Cathedral. In Structural Analysis of Historical Constructions II.

Possibilities of Numerical and Experimental Techniques, Roca et al. (Eds.), CIMNE, Barcelona, 1998.

169. [RSS] Rank, E., Schweingeruber, M. and Sommer, M., Adaptive mesh generation and transformation of triangular to quadrilateral meshes. Communications in Numerical Methods in Engineering, 9, 121-129, 1993.

170. [Sa] Samuelsson, A., The global constant strain condition and the patch test.

171. Energy Methods in Finite Element Analysis, R. Glowinski, E.Y. Rodin and O.C. Zienkiewicz (Eds.), 3, 46-68, J. Wiley, 1979.

172. [SC] Sunder, K.S. and Cookson, R.A., Integration points for triangles and tetra-hedrons obtained from the Gaussian quadrature points for a line. Computers & Structures, 21(5), 881-885, 1985.

173. [Sch] Schwer, L.E., Guide for verification and validation in computational solid mechanics. IACM Expressions, N. 20, 14-21, January 2007 (www.iacm.info).

174. [SDH] Stein, E., De Borst, R. and Hughes, T.J.R. (Eds.), Encyclopedia of Com-putational Mechanics. 1 Fundamentals. J. Wiley, 2004.

175. [SDH2] Stein, E., De Borst, R. and Hughes, T.J.R. (Eds.), Encyclopedia of Computational Mechanics. 2 Solids and Structures. J. Wiley, 2004.

176. [SF] Strang, S. and Fix, G.J., An analysis of the finite element method. Prentice Hall, 1973.

177. [SFPC] Su¶arez, S., Foces, A., Pereda, J. and Cacho, M., Treating the geometry of a °oating caisson to obtain a structured mesh in GiD. 1st Conference on Advances and Applications of GiD, Publication CIMNE No. 210, Barcelona, 2002.

178. [SG] Shephard, M.S. and Georges, M.K., Automatic three-dimensional mesh generation by the finite octree technique. Int. J. Num. Meth. Engrg., 32, pp. 709-749, 1991.

179. [SG2] Suarez, B. and Gonzalez, J.M., Analysis of Melonares dam. CIMNE Re-port (in Spanish), Barcelona, 2008.

180. [SG3] Suarez, B. and Gonzalez, J.M., Structural analysis and design of an arch in Jaen (Spain) accounting for the effect of the terrain. CIMNE Report (in Spanish), Barcelona, 2008.

181. [SHM] Suarez, B., Hanganu, A. and Miquel, J., Optimum design of multicellular reinforced concrete box docks. Journal of Structural Engineering (ASCE),

182. 128(5), 603-611, 2002.

183. [SM] Suarez, B. and Martel, E., Finite element analysis of Santa Coloma dam

184. (in spanish). Pub. IT-26, CIMNE, Barcelona, 1991.

185. [SM2] Suarez, B., and Martel, E., Analysis of turbine component in a hydroelec-trical power plant. CIMNE, Report, Barcelona, 1991.

186. [SR] Stein, E. and RÄuter, M., Finite element methods for elasticity with error-controlled discretization and model adaptivity. Encyclopedia of Computa-tional Mechanics, Vol. 1, 5{58, E. Stein, R. de Borst and T.J.R. Hughes (Eds.), Wiley, 2004.

187. [SRO] Stein, E., RÄuter, M. and Ohnimus, S., Adaptive finite element analysis and modelling of solids and structures. Findings, problems and trends. Int.

188. J. Num. Meth. Engrg., 60, 103-138, 2004.

189. Schroeder, W. and Shephard, M.S., Geometry-based fully automatic mesh generation and the Delaunay triangulation. Int. J. Num. Meth, Engrg., 26, 2503-2524, 1988.

190. [SS2] Schroeder, W. and Shephard, M.S., Computational Visualization. In Ency-clopedia of Computational Mechanics. Chapter 18, 1 Fundamentals, E. Stein, R. De Borst, J.R. Thomas, editors. J. Wiley, 2004.

191. [Str] Strang, G., Linear Algebra and its Application. Academic Press, New York, 1976.

192. [Ta] Taig, O.C., Structural analysis by the matrix displacement method. Electric Aviation Report, No. 5017, 1961.

193. [Tay] Taylor, R.L., Solution of linear equations by a profile solver. Engineering Computations, 2, 344-350, 1985.

194. [TBM] Tautges, T.J., Blacker, T.D. and Mitchell, S.A., The Whisker-Weaving algorithm: A connectivity-based method for constructing all-hexahedral finite element meshes. Int. J. Num. Meth, Engrg., 39, 3327-3349, 1996.

195. [TBW] Taylor, R.L., Beresford, P.J. and Wilson, E.L., A non

conforming element for stress analysis. Int. J. Num. Meth. Engrg., 10, 1211-20, 1976.

196. [TCMT] Turner, M.J., Clough, R.W., Martin, H. y Topp, L.J., Stiffness and defection analysis of complex structures. J. Aeron. Sci., 23, 805-23, 1956.

197. [TG] Timoshenko, S.P. and Goodier, J.N., Theory of elasticity. McGraw-Hill, New York, 3rd Edition, 1969.

198. [Ti] Timoshenko, S.P., Strength of Materials. Part 1: Elementary Theory and Problems. Van Nostrand Company Inc., New York, 1958.

199. [Ti2] Timoshenko, S.P., Strength of Materials. Part 2: Advanced Theory and Problems. 3rd. Edition, Van Nostrand-Reinhold, New York, 1956.

200. [TK] Taig, I.C. and Kerr, R.I., Some problems in the discrete element represen-tation of aircraft structures. In Matrix Methods of Structural Analysis, B.M. Fraeijs de Veubeke (ed.), Pergamon Press, 1964.

201. [TK2] Topping, B.H.V. and Khan, A.I., Parallel Finite Element Computations. Saxe-Coburg Publications, Edinburgh, UK, 1996.

202. [TMIP+] Topping, B.H.V., Muylle, J., Ivanyi, P., Putanowicz, R. and Cheng, B., Finite Element Mesh Generation. Saxe-Coburg Publications, Stirling, 2004.

203. [TSW] Thompson, J.F., Soni, B.K. and Weatherill, N.P. (Eds.), Handbook of Grid Generation. CRC Press, January 1999.

204. [TTM] Thompson, J.F., Thames, F.C. and Martin C.W., Automatic numerical generation of body-¯tted curvilinear coordinates for a field containing any number of arbitrary two dimensional bodies. J. Comput. Phys., 15, 299-319, 1974.

205. [TW] Thompson, J.F. and Warsi, Z.U.A., Boundary-fitted coordinate systems for numerical solution of partial differential equations. J. Comput. Phys., 47, 1-108, 1982.

206. [TWM] Thompson, J.F., Warsi, Z.U.A. and Martin C.W., Numerical Grid Gen-eration: Foundations and Applications. North-Holland, Dordrecht, 1987.

207. [TY] Timoshenko, S.P., and Young, D.M., Theory of Structures. 2d

Edition. McGraw-Hill, New York, 1965.

208. [Ut] Utku, S., Explicit expressions for triangular torus element stiffness matrix. AIAA Journal, 1174-1176, 1968.

209. [Ve] VerfÄurth, R., A review of a Posterior Error Construction and Adaptive Refinement Procedures. Wiley-Teubner, 1996.

210. [Vo] Voronoi, G., Nouvelles applications des paramµetres continus µa la theorie des formes quadratiques. J. Reine Angew. Math, 133, 97-178, 1907.

211. [vRBBB] van Rens, B.J.E., Brokken, D., Brekelmans, W.A.M. and Baaijens, F.P.T., A two dimensional paving mesh generator for triangles with con-trollable aspect ratio and quadrilaterals with high quality. Engineering with Computers, 14(3), 248-259, 1998.

212. [Was] Washizu, K., Variational methods in elasticity and plasticity. Pergamon Press, 1975.

213. [Wat] Watson, D.F., Computing the n-dimensional Delaunay tessellation with application to Voronoi polytopes. Comp. J., 24, 167-172, 1981.

214. [We] Weatherill, N.P., A method for generating irregular computation grids in multiply connected planar domains. Int. J. Num. Meth, Engrg., 8, 181-197, 1988.

215. [We2] Weatherill, N.P., The integrity of geometrical boundaries in the two-dimensional Delaunay triangulation. Commun. Appl. Numer. Methods, 6, 101-109, 1990.

216. [WH] Weatherill, N.P. and Hassan, O., Efficient 3-dimensional Delaunay trian-gulation with automatic point generation and imposed boundary constraints.Int. J. Num. Meth, Engrg., 37, 2005-2039, 1994.

217. [WJ] Weaver, W. (Jr.) and Johnston, P.R., Finite elements for structural analysis. Prentice Hall, 1984.

218. [WL] Wiberg, N.-E. and Li, X.D., Superconvergent patch recovery of finite element solutions and a posteriori l2 norm error estimate. Commun. Numer. Methods Engrg., 10, 313-320, 1994.

219. [WR] Wilkinson, J.H. and Reinsch, C., Linear Algebra. Handbook for Automatic Computation. Volume II. Springer-Verlag, Berlin, 1971.

220. [Ya] Yang, T.Y., Finite Element Structural Analysis. Prentice Hall, 1986.

221. [Yam] Yamaguchi, F., Curves and Surfaces in Computer Aided Geometric De-sign. Springer- Verlag, Berlin, 1988.

222. [YS] Yerry, M.A. and Shephard, M.S., Automatic three-dimensional mesh generation by the modified octree technique. Int. J. Num. Meth. Engrg., 20, 1965-1990, 1984.

223. [ZL] Zhang, Z. and Lin, R., Ultraconvergence of ZZ patch recovery at mesh symmetry points. Numer. Math, 95, 781-801, 2003.

224. [ZP] Zienkiewicz, O.C. and Phillips, D.V., An automatic mesh generation scheme for plane and curved surfaces by isoparametric coordinates. Int. J. Num. Meth, Engrg., 3, 519-528, 1971.

225. [ZS] Zienkiewicz, O.C. and Scott, F.C., On the principle of repeability and its application in the analysis of turbine and pump impellers. Int. J. Num. Meth. Engrg., 4(3), 445-448, 1972.

226. [ZT] Zienkiewicz, O.C. and Taylor, R.L. The Finite Element Method for Solid and Structural Mechanics. Sixth Edition, Elsevier, 2005.

227. [ZTZ] Zienkiewicz, O.C., Taylor, R.L. and Zhu, J.Z., The Finite Element Method. Its Basis and Fundamentals. Sixth Edition, Elsevier, 2005.

228. [ZVK] Zienkiewicz, O.C., Valliappan, S. and King, I.P., Stress analysis of rock as a no tension material. Geotechnique, 18, 56, 1968.

229. [ZZ] Zienkiewicz, O.C. and Zhu, J.Z. The superconvergent patch recovery (SPR) and a posteriori error estimations. Part I: The recovery technique. Int. J. Num. Meth. Engrg., 33(7), 1331-1364, 1992.

230. [ZZ2] Zienkiewicz, O.C. and Zhu, J.Z., The superconvergent patch recovery (SPR) and a posteriori error estimations. Part II: Error estimates and adaptivity. Int. J. Num. Meth. Engng., 33(7), 1365-1382, 1992.

231. [ZZ3] Zienkiewicz, O.C. and Zhu, J.Z., The superconvergent patch recovery (SPR) and and adaptive finite element re¯nement. Comput. Meth. in Appl. Mech. Engrg, 101, 207{224, 1992.

232. [ZZHW] Zhu, J.Z., Zienkiewicz, O.C., Hinton, E. and Wu, J., A new approach to the development of automatic quadrilateral mesh generation. Int. J. Num. Meth. Engrg., 32, 849-866. 1991.

INDEX

A

B

C

D

E

F

G

H

I

L

M

N

O

P

S

T

V

W